44th Symposium on Nuclear Physics Cocoyoc 2023

Journal of Physics: Conference Series Volume 2619

Cocoyoc, Mexico
9-12 January 2023

Editors:

Luis Acosta
P. Amador-Valenzuela
Roelof Bijker
D. J. Marín-Lámbarri

ISBN: 978-1-7138-8207-7
ISSN: 1742-6588

Printed from e-media with permission by:

Curran Associates, Inc.
57 Morehouse Lane
Red Hook, NY 12571

CURRAN ASSOCIATES INC.
proceedings
.com

Some format issues inherent in the e-media version may also appear in this print version.

Institute of Physics
Dirac House, Temple Back
Bristol BS1 6BE UK

Phone: 44 1 17 929 7481
Fax: 44 1 17 920 0979

techtracking@iop.org

Additional copies of this publication are available from:

Curran Associates, Inc.
57 Morehouse Lane
Red Hook, NY 12571 USA
Phone: 845-758-0400
Fax: 845-758-2633
Email: curran@proceedings.com
Web: www.proceedings.com

TABLE OF CONTENTS

Journal of Physics: Conference Series **2619** (2023) 011001 doi:10.1088/1742-6596/2619/1/011001

XLIV Symposium on Nuclear Physics

Hacienda Cocoyoc, Morelos, Mexico
January 9-12, 2023

Editors:

LUIS ACOSTA
Instituto de Física
Universidad Nacional Autónoma de México
PAULINA AMADOR-VALENZUELA
Instituto Nacional de Investigaciones Nucleares
ROELOF BIJKER
Instituto de Ciencias Nucleares
Universidad Nacional Autónoma de México
DANIEL JOSÉ MARÍN-LÁMBARRI
Instituto de Física
Universidad Nacional Autónoma de México

Journal of Physics: Conference Series **2619** (2023) 011001 doi:10.1088/1742-6596/2619/1/011001

International Advisory Committee

Daniel Bardayan	University of Notre Dame, USA
Francesco Cappuzzello	INFN-Laboratori Nazionali del Sud, Italia
Osvaldo Civitarese	Universidad Nacional de La Plata, Argentina
Martin Freer	University of Birmingham, UK
Alfredo Galindo-Uribarri	Oak Ridge National Laboratory, USA
María José García Borge	CSIC-Madrid, Spain
David Jenkins	University of York, UK
Reiner Krücken	TRIUMF, Canada
Jorge López	University of Texas at El Paso, USA
Thomas Papenbrock	University of Tennessee, USA
Jorge Piekarewicz	Florida State University, USA
Norbert Pietralla	T.U. Darmstadt, Germany
Rosario Gianluca Pizzone	INFN-Laboratori Nazionali del Sud, Italia
W. Michael Snow	Indiana University, USA
Adam Szczepaniak	Indiana University, USA
Michael Wiescher	University of Notre Dame, USA

Organizing Committee

Luis Acosta	Instituto de Física, UNAM
Paulina Amador-Valenzuela	Instituto Nacional de Investigaciones Nucleares
Roelof Bijker	Instituto de Ciencias Nucleares, UNAM
Daniel José Marín-Lámbarri	Instituto de Física, UNAM

Sponsors

Coordinación de la Investigación Científica, UNAM
Instituto de Ciencias Nucleares, UNAM
División de Física Nuclear de la Sociedad Mexicana de Física (DFN-SMF)

Journal of Physics: Conference Series **2619** (2023) 011001 doi:10.1088/1742-6596/2619/1/011001

Photograph of the event

Journal of Physics: Conference Series **2619** (2023) 011001 doi:10.1088/1742-6596/2619/1/011001

Preface

The 44th Symposium on Nuclear Physics was celebrated from January 9th to 12th, 2023 at the Hacienda Cocoyoc, Morelos, Mexico. This Symposium has been held every year since its inauguration in 1978 until 2020 despite earth quakes, last minute cancelations by the venue and economic crises. The COVID-19 crisis presented too much of a challenge and the Symposium had to be postponed for two years in a row.
In this 2023, the Symposium is back! We had a tremendous response from the international community of nuclear physics which led to a very successful and lively meeting. The program consisted of 32 plenary talks and 20 posters over a broad range of hot topics, such as nuclear structure and reactions, instrumentation, relativistic heavy ions, nuclear astrophysics, fundamental symmetries and hadron physics.
This proceedings contain 20 articles of exciting new results and developments in progress of amazing instrumentation that were presented at the meeting.
At the meeting there were representatives of several international collaborations to report on the most recent research progress and the most recent developments in laboratories like FRIBS, ORNL, INFN-LNS, INFN-LNL, INFN-LUNA, ININ and the Texas A&M, Notre Dame University and IF-UNAM facilities. During the presentations, recent data from NUMEN, AGATA, SUGAR@LNL, NNBAR, nEDM@SNS and the NOPTREX collaborations were shown. Many apparatus were as well presented, either as recent developments or showing their recent results; such as the K600 spectrometer, LHASA, SUGAR, AGATA, MUGAST, VAMOS, TriSol, FRAISE, FARCOS, CHIMERA and PROSPECT-I.
Regarding the theoretical studies, interesting new developments involving different topics were presented on this edition, such as Shell Model, QCD, QED, Ab-initio, baryons and pentaquarks, Nuclear Equation of State, Artificial Intelligence Approaches, nuclear clusters, R-matrix, nuclear dynamics and neutrino physics. The high quality of the talks and posters, the expertise of the speakers and the broad spectrum of subjects covered in the meeting, shows that nuclear physics continued working hard despite the Pandemic restrictions, being a very active area at the frontier of scientific research which establishes bridges between many different disciplines, putting together theoretical and experimental researchers and having an extension to the instrumentation developments and applications.
Unfortunately, during the last three years we had to say goodbye to four distinguished colleagues who made pioneering contributions to nuclear physics in Mexico and were very instrumental in the success of this series of symposia. María Esther Ortiz (1936-2021) was the last surviving member of the organizing committee of the first symposium in 1978, Jorge Flores (1941-2020) and Pier Mello (1939-2022) were among the first graduate students in nuclear theory in Mexico, and Alfonso Mondragón (1941-2020) was a regular participant in the Symposium for some 35 years. All four were members of the Instituto de Física of the UNAM. The 44th Symposium is dedicated to their memory in celebration of their distinguished careers and in recognition of their important contributions to the development of nuclear physics in Mexico. Some local colleagues presented part of the history of these distinguish researchers at the beginning of the Symposium. In the following lines, some words coming from this homage:

Journal of Physics: Conference Series **2619** (2023) 011001 doi:10.1088/1742-6596/2619/1/011001

Jorge Flores
(1941-2020)
(*by Efraín Chávez, IF-UNAM*)

Jorge Flores was awarded the PhD diploma in theoretical Nuclear Physics at the young age of 24, in times where nuclear science and technology were in everybody's mind: bombs, reactors and everyday more and more applications seeing the light of day. In his theses he used group theory to understand the structure of nuclei in the s-d shell.

Among the many awards he received during his career it is worth mentioning: The National Academy of Sciences (1972), The National Autonomous University of Mexico in Sciences (1988) and the "Kalinga" UNESCO (1992).

He became the youngest director of the Instituto de Física, at age 33 (1974-82). In this period, he co-founded the "Oaxtepec (today Cocoyoc) Nuclear Physics Symposium" in 1978. He also gave us here multiple invited talks throughout his life.

His role in securing funds for the Symposia in the critical first years allowed this event to establish itself as one of the main nuclear physics meetings in the world. For decades now, the first days of the year see scientists from all over the world gather in the warm lands of Oaxtepec (now Cocoyoc) to discuss the newest discoveries and progress in Nuclear Science.

Thank you so very much Jorge!

Journal of Physics: Conference Series **2619** (2023) 011001 doi:10.1088/1742-6596/2619/1/011001

Alfonso Mondragón
(1941-2020)
(*by Eduardo Peinado, IF-UNAM*)

Professor Alfonso Mondragón Ballesteros obtained a Physics degree in 1955 with a thesis written under Marcos Moshinsky's supervision. Afterward, under Rudolf Peierls supervision, he graduated from the University of Birmingham's Mathematical Physics department in 1960 with the thesis "The optical model of nuclear reactions with Pauli principle".

His compromise with the Symposium on Nuclear Physics was solid. He participated from the second to the 2014, only missing it in two editions, and organized the 1999 edition. I remember vividly that during my time as a graduate student, "el profe" attended almost every year the Cocoyoc workshop, presenting our results on flavor problems in the Standard Model in Cocoyoc.

Alfonso was a theoretical physicist who always tried to keep in touch with experimentalists, which he emphasized to his students. I remember he invited his experimental colleague and friend, P. Von Brentano.

Thank you very much, Profe!

Journal of Physics: Conference Series **2619** (2023) 011001 doi:10.1088/1742-6596/2619/1/011001

María Esther Ortiz
(1936-2021)
(by Libertad Barrón, IF-UNAM)

María Esther Ortiz y Salazar, aware of the fact that Physics is made in team, was a community builder. Solidary and generous, she dedicated a lot of her time and effort to serve in different committees that contributed to the development of Physics in Mexico. At the National Autonomous University of Mexico (UNAM), she was a Member of the Governing Board, Head of the Department of Experimental Physics at the Institute of Physics and Physics Coordinator at the School of Sciences.
She served as President of the Mexican Physical Society as well as President of its Nuclear Physics Division. Also Head of the Council for the Certification of Educative Programs in Physics.
Her contribution to the development of infrastructure for Nuclear Physics in Mexico was crucial for the facilities we have now, and also had a determining role in the consolidation of the Symposium on Nuclear Physics of the Mexican Physical Society.
A woman ahead of her time, a feminist, she was able to resist the impulse that most female physicist experience, sooner or later, to mimetize. She never stopped being herself to be a physicist, and with that inspired several generations of women in physics.

Journal of Physics: Conference Series **2619** (2023) 011001 doi:10.1088/1742-6596/2619/1/011001

Pier Mello
(1936-2021)

(by Manuel Torres, IF-UNAM)

Pier Achille Mello Picco was emeritus researcher of the Theoretical Department at IFUNAM since 1994. Regarding nuclear Physics, he studied the statistical theory of the nuclear reactions from a maximum entropy approach.

He studied Physics at the Sciences Faculty, UNAM, obtaining his degree with a thesis on theoretical models of the atomic nucleus, directed by Dr. Marcos Moshinsky, with whom he collaborated until his doctorate. In addition, he did postdoctoral work at Princeton University, where start to be interested in the theory of nuclear reactions. Once he was back in Mexico, he began a collaboration with Tomás Brody, together they published an article on stochastic matrices which has received more than a thousand citations.

Among the many prizes to his trajectory, he obtained the Marcos Moshinsky medal in 2005 and the Recognition of National Emeritus Researcher by the National Council of Science and Technology in 2006.

He was director of the Revista Mexicana de Física during the period of 1975-1977.

****

Finally, we would like to thank all our colleagues for their enthusiastic participation in this 44th edition of the meeting. In particular, we are grateful to the members of the International Advisory Committee for their invaluable suggestions and to the anonymous referees for their time and dedication in helping us to maintain the high standard of the contributions published in the proceedings of the Symposium.

Luis Acosta
Paulina Amador-Valenzuela
Roelof Bijker
Daniel José Marín-Lámbarri

Journal of Physics: Conference Series **2619** (2023) 011002 doi:10.1088/1742-6596/2619/1/011002

Peer Review Statement

All papers published in this volume have been reviewed through processes administered by the Editors. Reviews were conducted by expert referees to the professional and scientific standards expected of a proceedings journal published by IOP Publishing.

- **Type of peer review:** Single Anonymous
- **Conference submission management system:** Morressier
- **Number of submissions received:** 21
- **Number of submissions sent for review:** 21
- **Number of submissions accepted:** 20
- **Acceptance Rate (Submissions Accepted / Submissions Received × 100):** 95.2
- **Average number of reviews per paper:** 1
- **Total number of reviewers involved:** 16
- **Contact person for queries:**
  Name: Luis Acosta
  Email: acosta@fisica.unam.mx
  Affiliation: Instituto de Física, UNAM

Journal of Physics: Conference Series **2619** (2023) 012001 doi:10.1088/1742-6596/2619/1/012001

Threshold anomalies in low-energy reactions with loosely-bound projectiles

E F Aguilera*[,1], F Torabi[2], J C Morales-Rivera[3]

[1] Departamento de Aceleradores y Estudio de Materiales, Instituto Nacional de Investigaciones Nucleares, Apartado Postal 18-1027, C. P. 11801, México, D. F., México
[2] Departamento de Física, Instituto Tecnológico de Aeronáutica, DCTA, 12228-900, São José dos Campos, SP, Brazil
[3] Facultad de Ciencias, Universidad Autónoma del Estado de México, Instituto Literario 100, Código Postal 50000, Toluca, México

E-mail: *eli.aguilera@inin.gob.mx

Abstract. Recent results concerning data for fusion and elastic scattering of low-energy neutron-halo ^{6}He and weakly-bound ^{6}Li projectiles, colliding with medium mass targets, are briefly reviewed. The extended optical model (EOM) was used, which includes two complex polarization potentials, one associated to fusion couplings and the other one associated to direct reaction couplings. Four systems having ^{6}Li projectiles and one with ^{6}He were analyzed. The extracted threshold anomalies are discussed.

1. Introduction

Much interest within the nuclear physics community has been generated by the study of reactions with loosely bound nuclei, stable or radioactive [1, 2]. Among the stable weakly bound nuclei, the most studied one is perhaps ^{6}Li, which in addition maintains a strong current interest [3–8]. As for radioactive projectiles, reactions induced by the neutron-halo nucleus ^{6}He have proved to be a most fruitful area of research for over two decades [2, 9–11].

The aim of the present work is to describe some interesting threshold anomalies that have been shown recently in reactions induced by low-energy projectiles of weakly-bound ^{6}Li and neutron-halo ^{6}He. Energies near or below the Coulomb barrier have been analyzed. The term threshold anomaly (TA) [12, 13] refers to a peculiar energy behavior of the corresponding optical model polarization potentials. This is usually associated to the normal closing down of direct reaction channels as the energy approaches the Coulomb barrier, which causes the potential strength to start decreasing at an energy around the barrier and eventually go to zero at even lower energies, when all reaction channels are finally closed. However, for the case of weakly-bound projectiles some direct-reaction channels may remain open even below the Coulomb barrier, which has led to the observation of a different kind of threshold anomaly, the so called breakup threshold anomaly (BTA) [14, 15]. The intention of the present work is to present a summary of some recent findings within this context.

Optical model potentials have been obtained [3, 11] which can describe simultaneously both the elastic scattering data and the fusion data reported for a given system. All systems analyzed correspond to medium-mass targets. For ^{6}Li projectiles, targets of ^{58}Ni, ^{59}Co, ^{64}Ni, ^{64}Zn have been considered [3] while for the halo projectile only the ^{64}Zn target was analyzed [11]. The formalism known as Extended Optical Model (EOM) [16,17] was used, where the potential is given by

$$U_{TOT} = V_{bare} + V_{Coul} - [iW_{int} + U_F + U_D]. \tag{1}$$

U_{TOT} includes a real nuclear part V_{bare} as well as the Coulomb potential V_{Coul} . It considers also a purely imaginary potential interior to the barrier W_{int} , as well as two complex polarization potentials, U_F and U_D, the first one associated to fusion couplings and the latest one associated to direct reaction couplings. One very interesting feature of this model is that, once the terms of the potential are determined, it is possible to calculate not only the total reaction cross section, in the usual manner, but also the fusion cross section can be calculated, through the respective absorption in the interior and the fusion polarization potentials. The corresponding expression is [16]

$$\sigma_{fus} = \frac{2}{\bar{h}v} < \chi^{(+)}|W_{int} + W_F|\chi^{(+)} > \quad , \tag{2}$$

where v is the relative velocity and $\chi^{(+)}$ is the distorted wave function that satisfies the Schrödinger equation corresponding to the full optical potential of Eq. 1.

The bare potential, V_{bare}, was taken as the Sao Paulo Potential (SPP) [18], which has no fitting parameters, and for the Coulomb potential V_{Coul}, a radius of 1.2 fm was used; W_{int} is a volume Woods-Saxon potential interior to the barrier, with depth, radius and diffuseness of 50 MeV, 1.0 fm and 0.2 fm, respectively. U_F ($V_F + iW_F$) and U_D ($V_D + iW_D$) are complex polarization potentials that can be associated to fusion and direct couplings, respectively.

For V_F and W_F a volume Woods-Saxon form was used with common radius and diffuseness (r_F = 1.4 fm, a_F = 0.43 fm). The respective strengths (V_{F0}, W_{F0}) were taken as fitting parameters. The direct polarization potentials were surface potentials, taken as derivative Woods-Saxon form, whose strengths (V_{D0}, W_{D0}) were fitting parameters. Both the real and the imaginary parts of U_D were supposed to have common values of radius and diffuseness, independent of energy, whose actual values for each system were given in Refs. [3, 11], respectively. For the case of the ^{6}Li + (^{58}Ni, ^{59}Co, ^{64}Ni, ^{64}Zn) systems, the values of (r_D, a_D) were (1.61, 0.68), (1.57, 0.84), (1.57, 0.91), and (1.60, 0.85) fm, respectively, while for ^{6}He + ^{64}Zn the corresponding values were (1.51, 1.22) fm.

In summary, the real and imaginary strengths of both the Direct and the Fusion polarization potentials are the only free parameters, so there are 4 free parameters. These parameters were varied to obtain a simultaneous fit to the elastic scattering angular distribution and the respective fusion cross section for each measured energy, always making sure about the validity of the Dispersion Relation (DR) (see Refs. [3,11]) for the obtained potentials.

2. Results for ^{6}Li projectiles

Fig. 1 shows the elastic scattering angular distributions for the four systems having ^{6}Li projectiles, along with the respective predictions from EOM. The data are well reproduced for all systems.

The reported fusion cross sections, represented by the black circles in Fig. 2, are mostly well reproduced by the calculations indicated by the red diamonds; but there are some exceptions, for the two lowest energy points of ^{6}Li + ^{58}Ni and ^{6}Li + ^{64}Zn. These points are underpredicted by the calculations. This underprediction is a consequence of forcing the fusion polarization potentials to satisfy the DR. However, if this condition is relaxed, one gets highly unphysical fusion polarization potentials [3]. Our hypothesis was that the data may have been originally

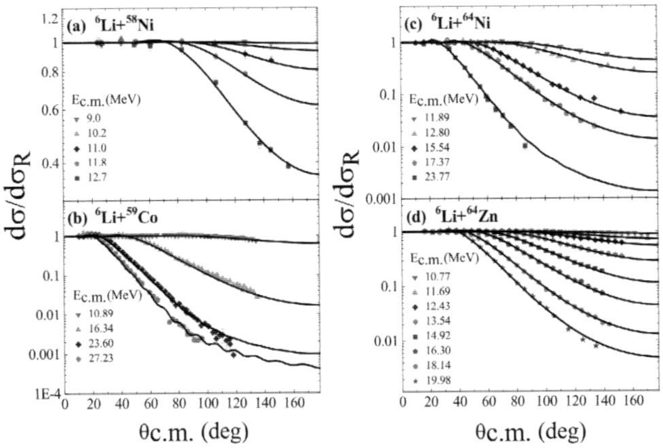

Figure 1. Elastic scattering angular distributions for ^{6}Li + (^{58}Ni, ^{59}Co, ^{64}Ni, ^{64}Zn). The data are from Refs. [19–22], respectively, and the curves represent the predictions of the EOM.

overpredicted, which is quite feasible because they were measured as inclusive data. Exclusive measurements are needed to corroborate this quite plausible hypothesis.

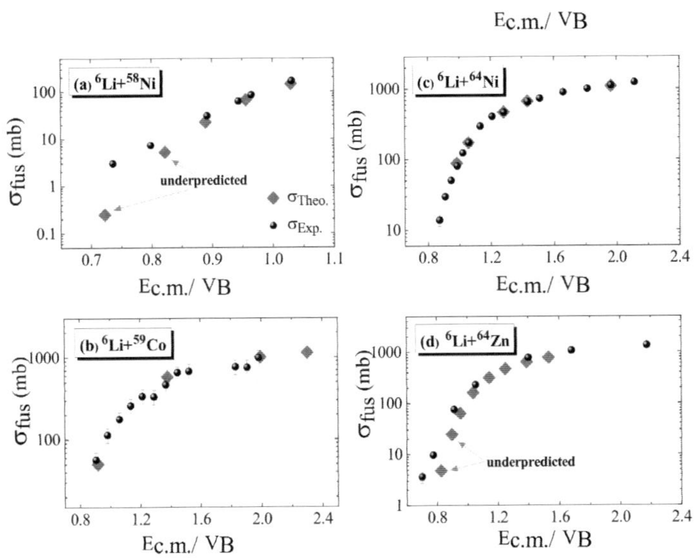

Figure 2. Measured and predicted fusion cross sections for ^{6}Li + (^{58}Ni, ^{59}Co, ^{64}Ni, ^{64}Zn). The data are from Refs. [23–26], respectively, and the red diamonds represent the predictions of the EOM.

Fig. 3 shows the strengths obtained for the real (red circles) and imaginary (black squares) polarization potentials in the case of ^{6}Li projectiles. The validity of the Dispersion Relation was always checked by approximating the imaginary parts by straight line segments which were used in the integral of the DR to obtain the predicted real parts, indicated by the dashed curves. The good consistency of these curves with the points extracted from the optimization corroborates the validity of the DR.

In the case of the points enclosed by the blue rectangles, for ^{6}Li + ^{58}Ni and ^{6}Li + ^{64}Zn, these points were forced to satisfy the Dispersion Relation, as mentioned above, causing an underprediction of the respective fusion cross sections. However, there are good reasons to

Journal of Physics: Conference Series **2619** (2023) 012001 doi:10.1088/1742-6596/2619/1/012001

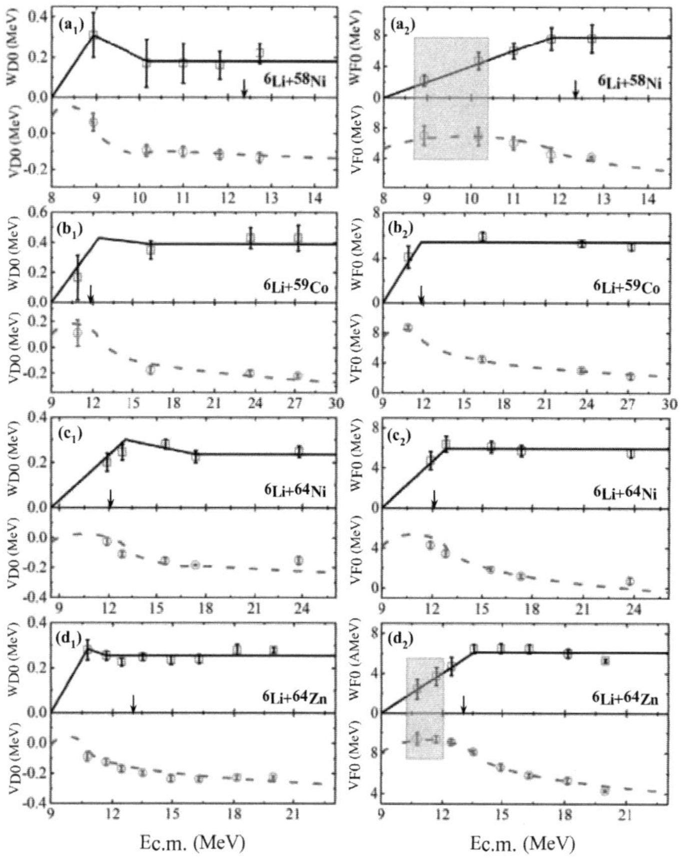

Figure 3. Strengths obtained for the direct (left hand side) and the fusion (right hand side) polarization potentials. The imaginary parts are approximated by the continuous lines and the dashed lines correspond to respective Dispersion Relation calculations. The blue rectangles are explained in the text.

believe that this procedure is correct.

It can be pointed out that the direct polarization potentials (left panel) are all consistent with the Breakup Threshold Anomaly. One can see a tendency for the W_D to increase with decreasing energy for energies around or below the barrier. Also, the real parts, V_D, are mostly repulsive (negative) for all systems. These features characterize the BTA.

On the other hand, the real parts of the fusion polarization potentials, V_F, are attractive and the imaginary parts, W_F, start decreasing around the barrier for all systems. This means that the fusion polarization potentials are all consistent with the TA rather than the BTA.

3. Results for ^{6}He projectiles

Switching now to the neutron-halo system, ^{6}He + ^{64}Zn, Fig. 4 presents the respective elastic scattering angular distributions. The curves describe the results of the EOM fits, which reproduce the data quite well.

In figure 5, the black filled circles represent the experimental fusion cross sections and the red empty circles refer to the respective calculated results. Considering the reported error bars, one can say that the calculations reproduce the data reasonably well. The blue empty squares represent the total reaction cross sections, calculated from the fitted potentials.

An interesting feature is that such empty squares show a local maximum, at 12.3 MeV,

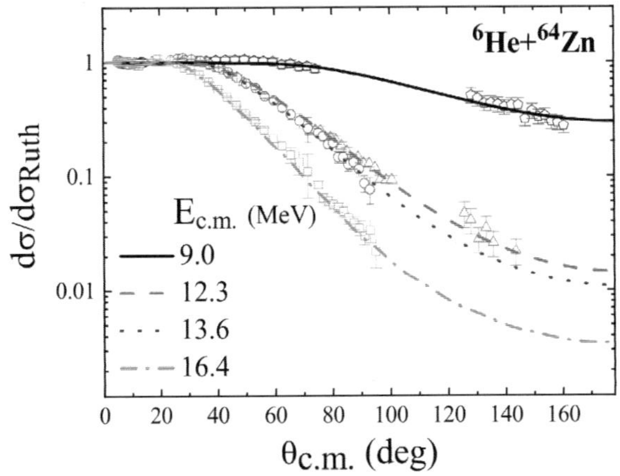

Figure 4. Elastic scattering angular distributions for ^{6}He + ^{64}Zn. Data are from Refs. [10, 27] and the curves represent EOM calculations.

which has been remarked in Fig. 5 by enclosing the involved points within an ellipse. This is unexpected, because total reaction cross sections normally grow larger with increasing energy. One can notice that the fusion cross sections don't show such a local maximum, indicating that it must originate from the respective direct reaction mechanisms. This will be corroborated below by looking at the behavior of the respective threshold anomalies.

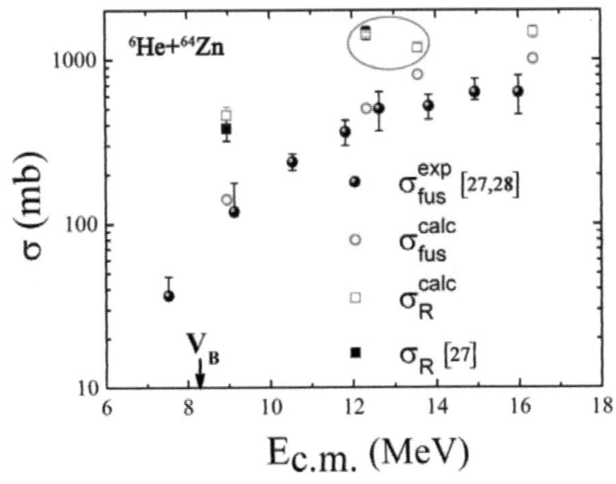

Figure 5. Fusion and total reaction cross sections for ^{6}He + ^{64}Zn. Data (filled symbols) are from Refs. [27, 28] and the empty symbols represent EOM calculations. The ellipse emphasizes the presence of a local maximum in the total reaction cross sections.

The right hand side panel of Fig. 6 shows the real and imaginary strengths of the direct reaction polarization potentials obtained for ^{6}He + ^{64}Zn. The imaginary strengths are approximated by the straight line segments, which are used in the Dispersion Relation to calculate the corresponding real part, given by the dashed red curves. One can see that the DR is well satisfied.

A strikingly irregular BTA is observed in this case. It is striking because of two reasons. On one hand, at the peak the strength rises quite high above the value corresponding to the next

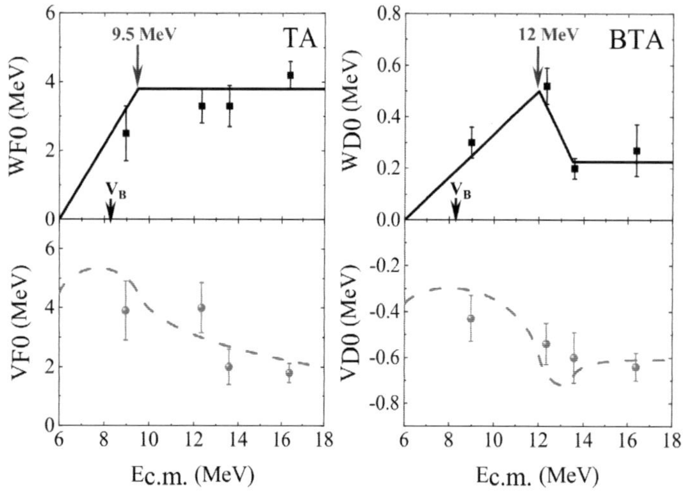

Figure 6. Potential strengths obtained for ^{6}He + ^{64}Zn. Continuous curves approximate the imaginary parts while dashed curves represent corresponding calculations for the respective real parts, using the Dispersion Relation.

higher energy. On the other hand, the energy at which this peak occurs is quite high above the barrier, in contrast to the situation for the ^{6}Li systems, where the peak appeared at energies close or even below the barrier.

The peak in this case is actually correlated with the local maximum in the total reaction cross sections pointed out previously, which also occurred at about 12 MeV. The conclusion is that there is some kind of transition which occurs in the involved direct reaction mechanisms, at an energy close to 12 MeV.

The left panel of Fig. 6 shows the strengths obtained for the fusion polarization potential, the imaginary part being on top and the real part on the bottom. Once again, one can see that the results satisfy the Dispersion Relation and, in this case, the typical behavior of the normal threshold anomaly is observed.

A possible mechanism explaining the results can be sketched. The ^{6}He can be thought of as formed by an alpha core and a 2-neutron halo. The Coulomb force due to the charge of the target tends to keep the alpha particle away from it but it doesn't act on the neutrons of the halo, so these neutrons could get close to the target and thus enhance the mutual nuclear attraction, which could lead to an increase in the total reaction cross section.

The probability that this situation occurs depends on the ratio between the relative speed of colliding nuclei (v_r) and the average speed of outer neutrons in the projectile (v_n). If this ratio is low, the neutrons can make several turns around the core during the interaction time, in which case the neutrons may have enough time to interact with the target and actually affect the reaction cross section. In the opposite situation, the effect of neutron rearrangement on the reaction cross section would be weaker or null.

The hypothesis is then that the observed local maximum in the reaction cross sections is the result of a transition between an adiabatic and a nonadiabatic rearrangement of the neutrons. It would be interesting to have real theoretical calculations which describe the time evolution of the loosely bound neutrons of ^{6}He in collisions with ^{64}Zn.

Additional phenomenological observations can be made, from a semiclassical point of view. By doing a transformation from angle to distance of closest approach on a Rutherford trajectory, a different representation of the angular distributions of Fig. 1 can be obtained. If the points

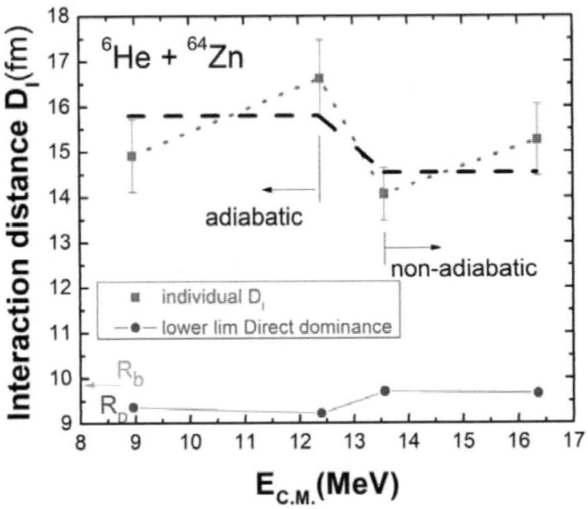

Figure 7. Interaction distance for ^{6}He + ^{64}Zn. For distances above R_D, W_D dominates over W_F, so mostly direct interactions occur. The barrier radius R_b is indicated for reference.

for all energies in a same system are plotted together, it can be shown that they follow a single trajectory [3]. This allows one to define a single interaction distance for each system, as the distance where the ratio to Rutherford starts decreasing from one [29–31].

For the case of the halo system, however, a large spread of points corresponding to the different energies appears in such a plot. Each energy must be analyzed individually in this case, which leads to different values of the interaction distance for each energy. Such values are plotted with red squares in Fig. 7, as a function of energy.

The transition mentioned above in direct reaction mechanisms is consistent with a transition from low to high values of the interaction distance when going from the assumedly non-adiabatic to the adiabatic region of energies. It is quite intuitive that in the adiabatic region, where the neutrons in the halo have more time to interact with the target, the interaction may occur at a larger distance. Measurements in an extended energy range would be desirable in order to further corroborate these results.

4. Conclusions

The results of EOM analyses for ^{6}Li + (^{58}Ni, ^{59}Co, ^{64}Ni, ^{64}Zn) and ^{6}He + ^{64}Zn are described. In all cases, the fusion part of the potential presented a normal Threshold Anomaly (TA), while the behavior of the direct reaction part was consistent with the Breakup Threshold Anomaly (BTA).

The Dispersion Relation was verified for all systems and it was pointed out that the constriction of having dispersive potentials is useful to identify seemingly unphysical data. For the particular case of the neutron-halo projectile, a strikingly irregular BTA was observed for the direct reaction part of the potential, which lead to the conclusion that an unexpected transition in the corresponding direct reaction mechanisms is present.

Acknowledgments

This work has been partially supported by CONACYT (México) under grant No. CB-01-254619.

References

[1] Canto L F, Gomes P R S, Donangelo R, Lubian J, Hussein M S 2015 *Phys. Rep.* **596** 1
[2] Kolata J J, Guimarães V and Aguilera E F 2016 *Eur. Phys. J.* A **52** 123

[3] Torabi F, Aguilera E F, Ghodsi O N, Gomez-Camacho A 2020 *Nucl. Phys.* A **994** 121661
[4] Amer A H, Penionzhkevich Yu E 2021 *Nucl. Phys.* A **1015** 122300
[5] Pakou A, Sgouros O, Soukeras V, Casal J, Rusek K 2022 *Eur. Phys. J.* A **58** 8
[6] Lubian J Ferreira J L, Rangel J, Cortes M R, Canto L F 2022 *Phys. Rev.* C **105** 054601
[7] Kumar R, Prajapat R, Maiti M 2023 *J. Phys.* G: Nucl. Part. Phys. **50**, 025106
[8] Alharbi W R, Ibraheem A A 2023 *Int. J. Mod. Phys.* E **32** 2350003
[9] Keeley N et al. 2009 *Prog. Part. Nucl. Phys.* bf 63 396
[10] Fernández-García J P, Di Pietro A, Figuera P, Gómez-Camacho J, Lattuada M, Lei J, Moro A M, Rodríguez-Gallardo M, and Scuderi V 2019 *Phys. Rev.* C **99** 054605
[11] Aguilera E F, Torabi F, and Morales-Rivera J C 2021 *Phys. Rev.* C **104** 054612
[12] Nagarajan M A, Mahaux C C, Satchler G R 1985 *Phys. Rev. Lett.* **54** 1136
[13] Mahaux C C, Ngô H, Satchler G R 1986 *Nucl. Phys.* A **449** 354
[14] Gomes P R S, *et al* 2005 *J. Phys.* G **31** S1669
[15] Hussein M S, Gomes P R S, Lubian J, and Chamon L C 2006 *Phys. Rev.* C **73** 044610
[16] Udagawa T and Tamura T 1984 *Phys. Rev.* C **29** 1922
[17] Udagawa T, Kim B T, and Tamura T 1985 *Phys. Rev.* C **32** 124
[18] Chamon L C, Carlson B V, Gasques L R, Pereira D, De Conti C, Alvarez M A G, Hussein M S, Cândido Ribeiro M A, Rossi Jr. E S and Silva C P 2002 *Phys. Rev.* C **66** 014610
[19] Aguilera E F, *et al* 2009 *Phys. Rev.* C **79** 021601(R)
[20] Souza F A, Leal L A S, Carlin N, Munhoz M G, Liguori Neto R, de Moura M M, Suaide A A P, Szanto E M, Szanto de Toledo A, and Takahashi J 2007 *Phys. Rev.* C **75** 044601; C. Beck, *et al*, arXiv:nucl-ex/0411002
[21] Biswas M, Subinit Roy, Sinha M, Pradhan M K, Mukherjee A, Basu P, Majumdar H, Ramachandran K, Shrivastava 2008 *Nucl. Phys.* A **802** 67
[22] Zadro M, *et al* 2009 *Phys. Rev.* C **80** 064610
[23] Aguilera E F, *et al* 2017 *Phys. Rev.* C **96** 024616
[24] Beck C, *et al* 2003 *Phys. Rev.* C **67** 054602
[25] Moin Shaikh Md, Subinit Roy, Rajbanshi S, Pradhan M K, Mukherjee A, Basu P, Pal S, Nanal V, Pillay R G, and Shrivastava A 2014 *Phys. Rev.* C **90** 024615
[26] Di Pietro A, *et al* 2013 *Phys. Rev.* C **87** 064614
[27] Di Pietro A, *et al* 2004 *Phys. Rev.* C **69** 044613
[28] Scuderi V., *et al* 2011 *Phys. Rev.* C **84** 64604
[29] Guimarães V, *et al* 2016 *Phys. Rev.* C **93** 064607
[30] Guimarães V, Lubian J, Kolata J J, Aguilera E F, Assuncão M, Morcelle V 2018 *Eur. Phys. J.* A **54** 223
[31] Guimarães V, Cardozo E N, Lubian J, Assuncão M, Pires K C C, Canto L F, Mukeru B, Kaur G, Aguilera E F 2021 *Eur. Phys. J.* A **57** 90

Journal of Physics: Conference Series **2619** (2023) 012002 doi:10.1088/1742-6596/2619/1/012002

Neutrino-Axion Couplings and the neutrino-mass

O. Civitarese[1,2]

[1] Department of Physics. University of La Plata. c.c.67 (1900) La Plata. Argentina
[2] Institute of Physics La Plata (IFLP). diag 113 and 63.(1900) La Plata, Argentina

E-mail: `osvaldo.civitarese@fisica.unlp.edu.ar`

Abstract. The Peccei and Quinn axions are low-mass pseudo-scalar neutral bosons whose existence was proposed to explain for the spontaneous breaking of the CP invariance in the early Universe. In the minimal $(SU(2)_{left} \otimes U(1))$ version of the leptonic sector of the electroweak lagrangian flavor neutrinos are massive. They are described as linear combinations of neutrino mass eigenstates and they oscillates between flavor states. However the mass mechanism which gives their masses is still unknown. In this talk we discuss a possible mechanism to explain for non-zero neutrino masses, which is based on the treatment of neutrino-axion interactions.

1. Introduction

The axion was proposed by R.D. Peccei and H.R. Quinn [1] as a possible solution to the CP problem. They have extended the Standard Model (SM) Lagrangian by adding a CP-violating term, and introduced a global axial U(1) symmetry with the axion: a neutral scalar complex field with a non-vanishing vacuum expectation value, as a mediator.

Accordingly to cosmological arguments, the Peccei-Quinn field acquired a non-zero vacuum expectation value $\langle\Phi\rangle_0$, leading to a spontaneous symmetry breaking at an unknown energy scale f_a [2]. To the vacuum expectation value of the axion field it can be associated a massive particle, the axion, with a mass m_a, through a Higgs-like mechanism, though the mass scale is not uniquely fixed, at variance with the standard Higgs scheme.

Among the various experiments devoted to the search of axions, we shall mention the ones of Refs.[3, 4, 5].

Axions do play a role in neutrino physics. In fact, the coupling of neutrinos with axions could provide a mechanism to explain for non-zero neutrino masses, as we shall explain. We shall briefly describe the neutrino-axion coupling sector of the Lagrangian and calculate the neutrino propagator at one loop. Some details of previous calculations which we have performed can be found in [6, 7].

2. Brief description of the formalism

2.1. The effective neutrino mass

The Lagrangian which describes the coupling between neutrinos and axions can be written as

$$\mathcal{L} = \imath \frac{g_a}{m_a} \bar{\psi}\gamma^\mu\gamma^5\psi\partial_\mu a \,, \tag{1}$$

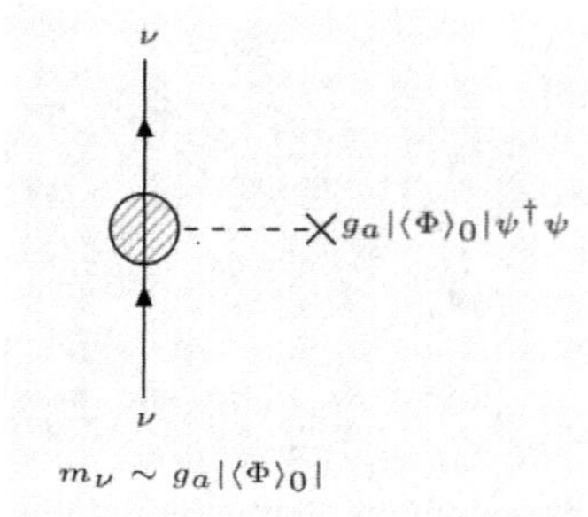

Figure 1. Mass insertion of the neutrino propagator due to the scalar coupling to the axions, in the symmetry-breaking regime.

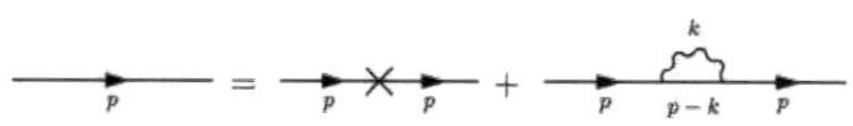

Figure 2. One loop corrections to the neutrino propagator

where the coupling constant g_a is dimensionless. The axion acquires a vacuum expectation value in presence of a potential

$$V(a) = -\frac{\mu^2}{2}\left(|a|^2 - \frac{|a|^4}{f_a^2}\right) . \tag{2}$$

Then, the variation of $V(a)$ leads to the breaking of the U(1) symmetry and gives the vacuum expectation value $< a >$

$$a \to (a(\bar{x}) + < a >)\, e^{-\imath m_a t} , \tag{3}$$

where $< a >= \frac{f_a}{\sqrt{2}}$, therefore Eq. (1) yields a neutrino mass term after performing the correspondence

$$m_\nu^{(0)} = g_a < a > , \tag{4}$$

This is given by the first order diagram, where the order refers to the power of the neutrino-axion coupling constant g_a, shown in Figure 1

To this, one adds the quadratic one-loop corrections, depicted in Figure 2.

The second order corrections to the neutrino propagator shown in Figure 2, is written

$$\imath S^{(2)}(p) = \imath S(p) + \imath S(p)\,(-\imath\Sigma\,(p)\,1)\,\imath S(p) , \tag{5}$$

where

$$\Sigma(p) = \int \frac{d^4k}{(2\pi)^4} \imath g_a \gamma^\mu \frac{\imath}{p-k-m} \frac{g_{\mu\nu}}{k^2 - m_a^2} \gamma^\nu \imath g_a , \tag{6}$$

 2619 (2023) 012002 doi:10.1088/1742-6596/2619/1/012002

p and k are the neutrino and the axion 4-momenta, m is the neutrino mass, m_a is the axion mass, and $g_{\mu\nu}$ is the metric. To compute the one-loop correction to the mass (Eq. 4) we shall work in $d = 4 + \epsilon$-dimensions and include a parameter ξ which has dimension of mass. The details of the calculations are presented in [8].

Taking $\Sigma(p)$ on-shell, the neutrino physical mass is written:

$$m_\nu \;=\; m + \Sigma(p)\Big|_{p^2=m^2} \,. \tag{7}$$

2.2. The neutrinoless double beta decay ($0\nu\beta\beta$)

The dependence of the neutrino mass with the axion mass and with the coupling between axions and neutrinos can be further restricted by the limits imposed on the neutrino mass by the non-observation of the neutrinoless double beta decay [9]. The $0\nu\beta\beta$ decay is a nuclear decay process which can take place only if the neutrino is a massive Majorana particle, identical to its antiparticle and/or if left-right and right-right terms are added to the Standard Model Lagrangian, implying the existence of right-handed $W_\pm$ and Z_0 mediators of the currents, with masses which should be larger than 2 TeV. It would also imply the existence of heavy mass neutrinos, with masses of the order of 100-200 MeV. The current limits for the half-life indicates that it should be larger than 10^{25} years and taken the contribution of the neutrino mass term and assuming furthermore average values for the nuclear matrix elements involved, the resulting electron neutrino mass should be smaller than 0.1 eV if the normal hierarchy is adopted for the neutrino mass-mixing sector, or smaller than 0.5 eV for the degenerate one. To be on the conservative side we shall take 0.5 eV as the upper limit and restrict the calculations of the axion-neutrino coupling by adopting this value. Therefore we can combine both results, the ones coming from the axion-neutrino coupling and those from the neutrinoless double beta decay, in order to reduce the number of free parameters involved in the calculations.

3. Results

3.1. Results for the effective neutrino mass

In Figure 3 we show the results of the calculations of the effective neutrino mass resulting from the coupling to axions. These values, which depend upon two parameters, like the mass of the axion and the coupling, should be taken as cuts in the (m_a,g_a) plane. For the chosen region the values of m_ν are limited by the non-observation of the neutrinoless double beta decay transitions [9].Also, we can proceed in opposite direction by fixing the neutrino mass and extracting from it the axion mass as a function of the coupling constant g_a. The results are shown in Figure 4.

4. Conclusions

In this talk we have presented the results obtained by coupling neutrinos to axions, as a possible mass mechanisms for the light neutrinos and combined the results with the limits extracted from the non-observation of the neutrinoless double beta decay. Both sets of results are compatible. It indicates that the existence of axions can indeed be taken as a reasonable working tool to explore the properties of the neutrino sector of the electroweak and particle channels related to it. More details about the formalism and calculations can be found in Refs.[6, 7, 8]

5. Acknowledgment

Some of the results presented in this talk are based on the work done in collaboration with M.E.Mosquera, from the Faculty of Astronomy and Geophysics of the National University of La Plata, to whom her contributions are gratefully acknowledged.

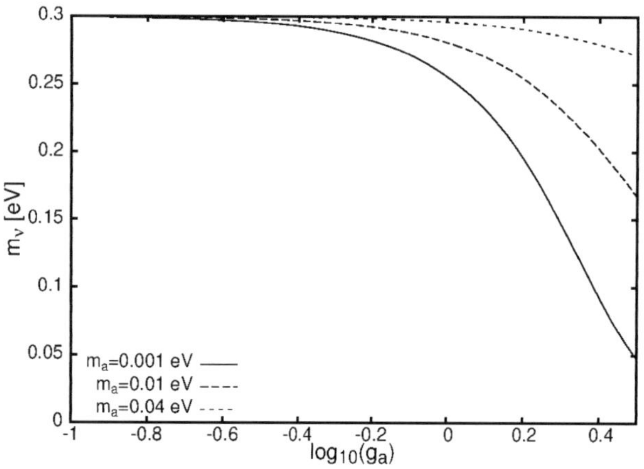

Figure 3. Mass of the neutrino, as a function of the axion-mass for different values of the axion-neutrino coupling constant.

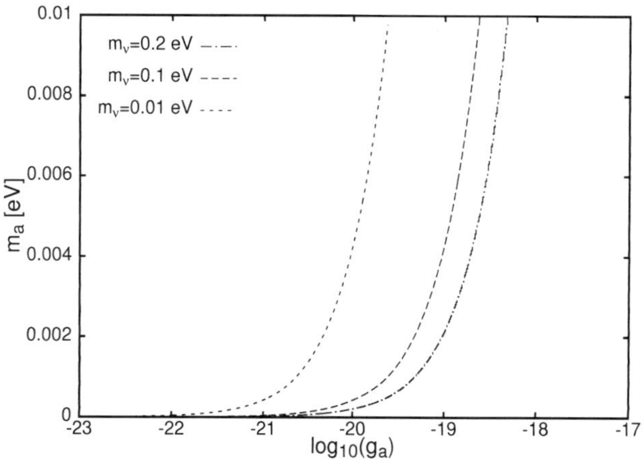

Figure 4. Mass of the axion, as a function of the neutrino-mass for different values of the axion-neutrino coupling constant.

6. References

[1] R. D.Peccei and H. R.Quinn. *Phys. Rev. Lett.* **38**, 1440 (1977)
[2] S. Weinberg. *The Quantum Theory of Fields. Vol II.* Cambridge University Press (1996).
[3] C. Bartram et al.*Phys. Rev. D.* **103**, 032002 (2021).
[4] A. Ejlli et al.*Phys. Rep.* **871**, (2020).
[5] A. Alvarez Melcon et al.*Journal of High Energy Physics* **75**, (2021).
[6] A. V. Penacchioni and O. Civitarese*Int.Journal Modern Physics E* **31**, 2250038 (2022)
[7] M. E. Mosquera and O. Civitarese*Int.Journal Modern Physics E* **31**, 2250082(2022)
[8] M. E. Mosquera and O. Civitarese*to be published*(2023)
[9] M. Agostini *et al.*, GERDA collaboration, *Phys. Rev. Lett.* **111**, 122503 (2013); A. Gando *et al.*, KamLAND-Zen collaboration, *Phys. Rev. Lett.* **110**, 062502 (2013); J. B. Albert *et al.*, EXO-200 collaboration, arXiv:1402.6956 (2014); see also Nature **510**,229 (2014).

Journal of Physics: Conference Series **2619** (2023) 012003 doi:10.1088/1742-6596/2619/1/012003

Ultra-peripheral nuclear collisions

Carlos A. Bertulani

Physics and Astronomy Department, Texas A&M University-Commerce, Commerce, Texas 75025, USA

E-mail: carlos.bertulani@tamuc.edu

Abstract. This article presents a very brief review of the physics of Ultra-Peripheral Collisions (UPC) at the Large Hadron Collider (LHC) and other nuclear facilities. I discuss several processes of interest such as electron-position pair production, the anti-hydrogen atom, giant resonances, exotic meson production and parton distribution functions.

1. Introduction

In 1985, my PhD adviser (Gerhard Baur, Jülich, Germany) and I came up with a paper on relativistic Coulomb excitation [1]. We showed that when the electromagnetic fields of one nucleus induces inelastic transitions in the other, the excitation process involves the same transition matrix elements as those induced by real photons. These collisions are known as *Ultra-Peripheral Collisions* (UPC). In first-order perturbation theory, a factorization can be carried out leading to a sum over multipoles [1],

$$\sigma = \sum_{E/M,L} \int d\omega \frac{n_{E/M,L}(\omega)}{\omega} \sigma_\gamma^{(E/M,L)}(\omega), \qquad (1)$$

where $\sigma_\gamma^{(E/M,L)}(\omega)$ are the cross sections induced by real photons with energy ω. The electric (E) and magnetic (M) multipolarities contain the photon angular momentum L component. The functions $n_{E/M,L}$ depend upon the projectile bombarding energy E_{beam} and the excitation energy ω. They are known as *equivalent (virtual) photon numbers* (EPN) [1]. At projectile energies below a few GeV/nucleon, the EPNs are strongly dependent on the multipolarity, e.g., $n_{E2} > n_{E1} > n_{M1}$ whereas at much large energies they are approximately equal, $n_{E2} \sim n_{E1} \sim n_{M1}$, except for the very low excitation energies $\omega \ll \gamma/b$ [1]. Here, $\gamma = (1-v^2)^{-1/2}$ is the Lorentz factor, v the relative velocity between the nuclei and b the impact parameter in the collision.

Enrico Fermi formulated a similar idea already in 1924 [2, 3] for the ionization of atoms by fast α-particles. His analytical result looks like Eq. (1), but is only valid for the E1 multipolarity. Fermi's method is known as *Weizsäcker-Williams (WW) method*, who introduced relativistic corrections in Fermi's formalism in 1934 [4, 5]. Intriguingly, Fermi published the same paper in two different journals in different languages (German in *Z. Phys.* and Italian in *N. Cimento*). Nobody really knows why, but one guesses that the reason is that Fermi was a young physicist looking for a faculty position in Italy and the version of his paper in *Nuovo Cimento* was intended to advertise his relatively unknown name in Italy at the time[1].

At the Large Hadron Collider (LHC) at CERN, Switzerland, the $\gamma\gamma$ collisions induced by the electromagnetic (EM) fields of the colliding ions occur at center of mass energies an order of magnitude higher than were available at now defunct e^+e^- colliders, and γ-nucleus collisions reach 30 times the energies available at fixed target accelerators. The Lorentz boost factor γ in the laboratory frame is about 7000 for p-p, 3000 for Pb-Pb collisions. Because of the large charge of Pb ions ($Z = 82$) and the short-interaction time, $\Delta t \simeq (10^{-20}/\gamma)$ seconds, the EM are stronger ($\propto Z^2$) than the Schwinger critical field [6, 7] $E_{Sch} = m^2/\hbar e = 1.3 \times 10^{16}$ V/cm. Light particles such as e^+e^--pairs are produced copiously by such fields [8].

This brief review provides an incomplete account of the development of UPCs since my PhD thesis was published in 1988 [8]. It is rewarding to see that, despite the skepticism reigning when my thesis was published, all the predictions we made in numerous works in the 1980s and 1990s have motivated a flurry of experimental results and theoretical developments. The physics of UPCs has become a subject of intensive studies worldwide and helps to consolidate numerous phenomena in QED and QCD [9, 10, 11, 12, 13, 14, 15].

2. Misconceptions

Because of its popularity in dealing with high energy processes, it is often assumed that the EPN formulated for UPCs is only valid in relativistic heavy ion collisions. That is wrong. Eq. (1) always holds if first-order perturbation theory is accurately enough to describe a low energy UPC process. This is a consequence of the assumption that during the excitation process $\nabla \times \mathbf{B} = 0$, $\nabla \cdot \mathbf{E} = 0$, where $\mathbf{E}$ and $\mathbf{M}$ are the EM fields generated by the source nucleus.

Another common misconception is that one cannot separate pure EM processes from those induced by the strong interaction. Numerous ways have been devised to separate the two processes. One often utilizes angular distributions, excitation energies, laboratory energy and target dependence of the cross sections, etc. In particular, measurements using light (e.g., carbon) concurrently with heavy (e.g., lead) nuclei helps to separate the contribution of the two interactions.

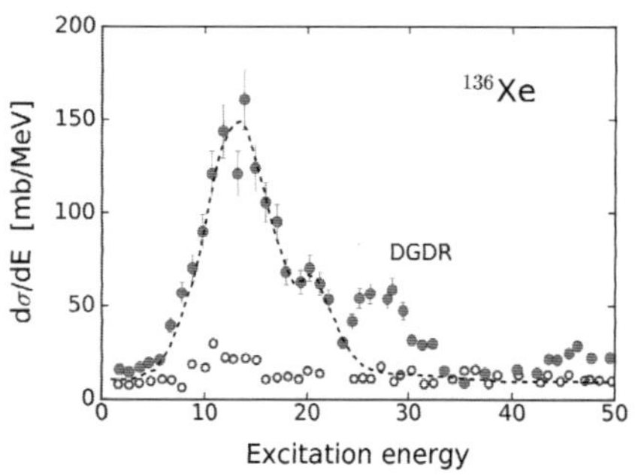

Figure 1. Excitation spectrum of giant resonances in ^{136}Xe in UPCs with a large-Z target. The DGDR was observed as a clear bump in the spectrum [25].

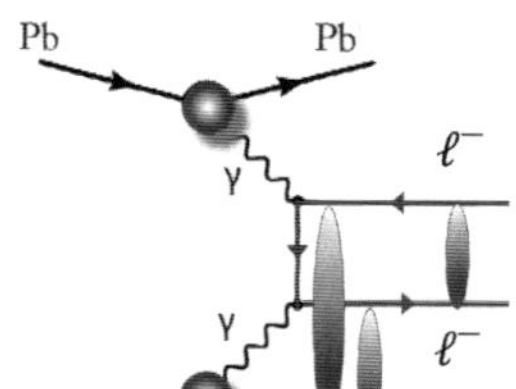

Figure 2. Lepton-antilepton pair production in UPCs. The blobs represent possible higher-order processes.

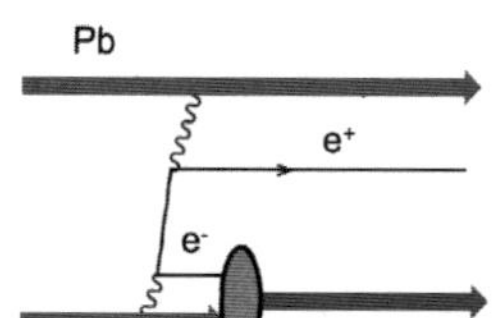

Figure 3. Pair production with electron capture leading to possible beam losses [22].

[1] Another possible explanation is that Nuovo Cimento, founded in 1923, in the beginning was not a real journal but collected reports from members of the Italian Physical Society (SIF) - I am indebted to Angela Bonaccorso for this clarification.

3. Giant Resonances

One of our first applications of UPCs was in the study of giant resonances in nuclei [16]. The excitation is most often followed by neutron emission and the cross sections are several barns for heavy nuclei even at moderate energies of $\sim$ 1 GeV/nucleon such as those available at the GSI laboratory in Germany [17, 18]. At these bombarding energies giant resonances can also be excited with a large probability by means of nuclear interactions. But, it was soon realized [8] that for large-Z nuclei the cross sections are much smaller than those induced by the EM interaction. Nowadays, Coulomb excitation and decay of giant resonances is a useful experimental probe, including nuclear fission studies [19, 20]. Due to its large cross section, the excitation and decay of giant resonances has been proposed as a luminosity monitor in heavy-ion colliders [21].

4. Double giant resonances

The large probabilities for excitation of giant resonances led to our proposal in 1986 [23, 24] that multiple giant resonances could also be excited via higher-order processes, or multiple photon exchange. When higher-order effects are of relevance, coupled-channels equations are one possible way to treat the process non-perturbatively. One can use the set of coupled-channel equations ($\hbar = c = 1$),

$$i\frac{d}{dz}\mathcal{A}_n(\mathbf{b}, z) = \sum_m \left\langle \Phi_n|\mathcal{O}_{E/M,L}(\mathbf{b}, z)|\Phi_m\right\rangle \mathcal{A}_m(\mathbf{b}, z)\; e^{i\omega_{nm}z}, \tag{2}$$

where $\mathcal{A}_n(\mathbf{b}, z)$ is the excitation amplitude for a collision at the impact parameter $\mathbf{b}$, z is the longitudinal coordinate, v is the projectile velocity, $\mathcal{O}_{E/M,L}$ is the appropriate electromagnetic operator, and n, m denote nuclear states. Diffraction effects induced by nuclear interactions can also be included with help of Glauber methods for nuclear absorption. We predicted large cross sections for the excitation of double, triple, and multi-phonon resonances in nuclei and proposed experimental measurements [23, 24], accomplished in two pioneer experiments at the GSI laboratory in Germany in 1993 [25, 26]. One experiment used gamma-gamma coincidences for the decay of the double giant dipole resonance (DGDR) [26] and another looked at its particle (neutron emission) decay [25].

In the 1980s, nuclear structure experts told us that it would be impossible to identify the DGDR in UPC experiments because it would be immersed in a large background of many-particle-many-hole excitations. History proved once again that there is nothing better than experimenting and observing. Figure 1 shows the results of the experiment reported in Ref. [25] for the DGDR excitation in ^{135}Xe projectiles in UPCs obtained at the GSI in 1993. Among the many reasons why the DGDR is worth studying is because its strength and width are not well understood and can be used to constrain nuclear models for the response to EM operators. Extensive discussions of these models have been published in Refs. [27, 28, 29].

5. Electron-positron pair production

Electron-positron pairs (Fig. 2) are copiously produced in UPCs. The first papers for the production of e^+e^- pairs in UPCs date back to the 1930s. Bethe, Racah, Bhabha, Tomonaga, Nishina, Furry, and many others, developed methods to calculate pair production using the newly born Dirac equation. Dirac predicted the existence of the positron when he formulated his famous equation for the description of the dynamics of electrons. The positron would manifest as a hole in the "vaccuum sea" of electrons. At the time the only way to look for this "hole" (positron) would be in UPCs with cosmic rays with large kinetic energies E. Assuming that the energy of the produced pairs are much larger than the electron rest mass, m_e, all the theoretical predictions yield a production cross section to leading order equal to $Z_1^2 Z_2 \alpha^2 \ln^3 \gamma$, where the Lorentz factor is $\gamma \simeq E/m_e$. In 1986, we revisited these calculations using Quantum Electrodynamics (QED), a practical formalism outside the reach of theorists in the 1930s. We have formulated a theory including final state interactions (the blobs in Fig. 2) using Bethe-Maximon distorted waves [8]. Because the cross sections are very large, of the order of *kilobarns* for a typical collider energy such as the LHC, we have shown that the role of higher-order corrections was worth studying [8, 30].

It is delightful to see that current experiments at the LHC are measuring the very fundamental pair-production cross sections first calculated in the 1930s. They are now almost fully understood theoretically [32] (see Fig. 4). During the first experiments, the predictions based on QED were confirmed (Fig. 4, adapted from Ref. [32])). The production of other particle-antiparticle pairs such as lepton-antilepton pairs (Figure 2), $\gamma\gamma \to \mu^+\mu^-$, $\gamma\gamma \to \pi^+\pi^-$, $\gamma\gamma \to W^+W^-$, etc, with such a mechanism is not negligible, as first found in Ref. [33, 8, 31, 22].

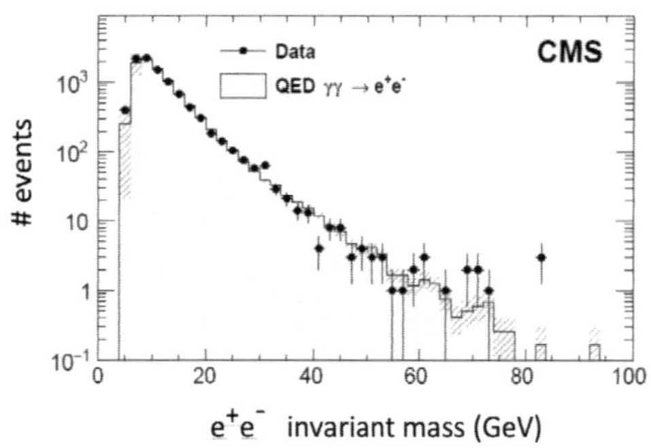

Figure 4. UPC production of e^+e^- pairs as observed in the CMS detector at the LHC (adapted from Ref. [32]).

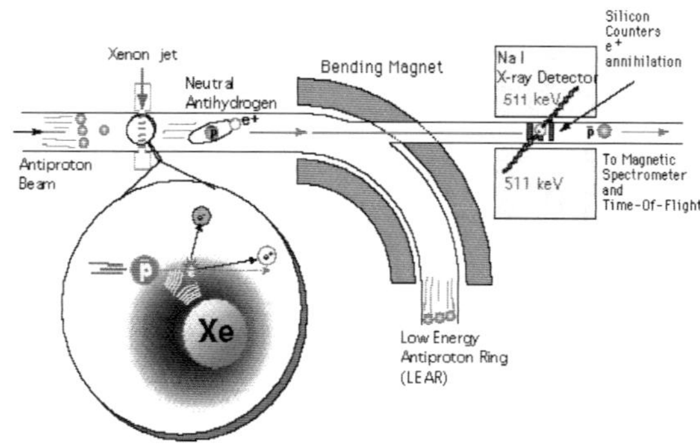

Figure 5. Schematic view of the setup used to identify anti-atoms at the LEAR/CERN in 1996 [35].

6. Anti-atoms

Arguably, the most beautiful application of UPCs is the production of anti-hydrogen via pair creation with the capture of the corresponding lepton by one of the colliding nuclei (Fig. 3) [8]. The application of this method to create anti-hydrogen atoms at the LHC was proposed in Ref. [34] and first observed by an ingenious experiment performed at CERN and published in 1996 [35]. At the Low Energy Antiproton Ring (LEAR) at CERN one used antiprotons colliding with protons and looked for the positron being captured in an atomic orbit around the antiproton. It was the first time anti-atoms were produced in a terrestrial laboratory with 11 anti-hydrogen atoms observed (see Fig. 5). The exciting news made headlines in newspapers around the world and were reported for the general public, such as in the front cover of the New York Times [36]. A few years later, a similar experiment was performed at the FERMILAB [37] with 57 events identified, consistent with our perturbative calculations performed before the experiment [38] (see also [39]). The properties of anti-atoms are now investigated using ion traps with the purpose to study fundamental symmetries [42, 43]. Larger antimatter atoms such as anti-deuterium, anti-tritium, and anti-helium can also be produced in UPCs [47]. Ref. [48] has made further predictions for the UPC production of muonic, pionic, and other exotic atoms by the coherent photon exchange between nuclei at the Large Hadron Collider (LHC). These predictions have not yet been observed.

It is also worth noticing that the process of electron-positron production with the capture of the electron in an atomic orbit of one of the nuclei was proposed as a source of beam loss in relativistic colliders already in 1988 in Ref. [31]. Our estimates at the time was a degradation of a large-Z ion beam within 2 hours at the LHC. This process was further studied [40] and recent experiments at the LHC [41] have confirmed our expectations [31]. Another process of interest is the production of ortho- and para-positronium in UPCs. Ref. [44] has provided a theoretical formalism based on quantum field theory to calculate positronium and other bound-states such as mesons (bound $q\bar{q}$) in $\gamma\gamma$- and $\gamma\gamma\gamma$-fusion in UPCs [45, 46].

7. Light-by-light scattering

The elastic scattering of light by light, $\gamma + \gamma \rightarrow \gamma + \gamma$, can only proceed via the fluctuation of a photon into an e^+e^- pair (Fig. 6). It has a rather small probability and has never been possible to study directly. In Ref. [49] we made the first proposal to use UPC to look for $Z_1 + Z_2 \rightarrow Z_1 + Z_2 + \gamma + \gamma$. The process would involve two virtual photons scattering via a box-diagram yielding two real photons as a byproduct. We have shown that the calculations based on the Delbrück scattering formalism was very uncertain theoretically [49]. This process was recently observed (2017) in an experiment at the LHC by the ATLAS collaboration [50] (Fig. 7). Such an observation opens the door for the search of physics beyond the Standard Model (SM) because a measured cross section larger than that predicted by the SM model [51] could be a signal of new particles such as axions [52, 53].

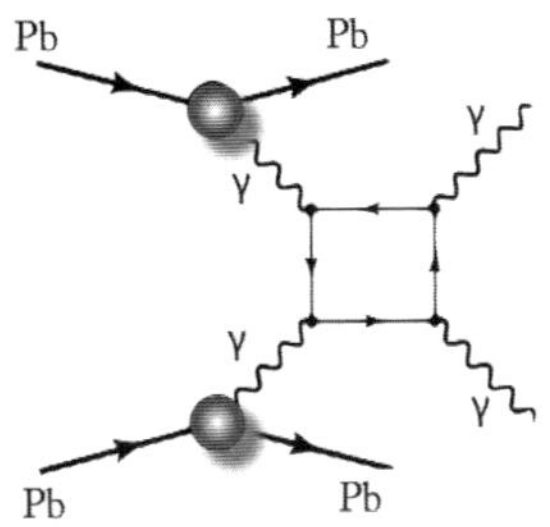

Figure 6. Feynman diagram for light-by-light scattering in UPCs.

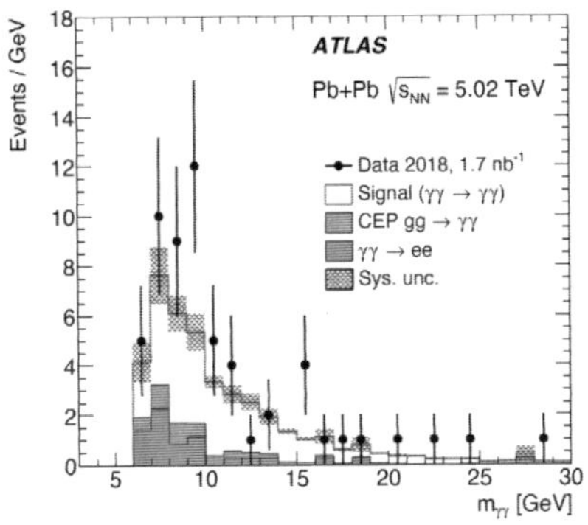

Figure 7. Light-by-light scattering events observed in UPCs at the ATLAS detector at the LHC [50].

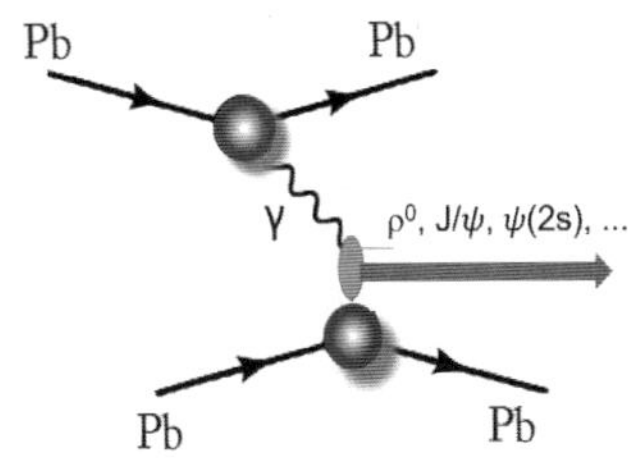

Figure 8. Schematic diagram for the production of vector mesons in UPCs.

8. Meson production in UPC

In Ref. [49] the EPN method was generalized to calculate the production of a bound-particle X in UPCs (Fig. 8). One can write the cross section for photon-photon fusion as [8],

$$\sigma_X = \int \frac{d\omega_1}{\omega_1}\frac{d\omega_2}{\omega_2} n_\gamma(\omega_1)\, n_\gamma(\omega_2)\, \sigma^X_{\gamma\gamma}(\omega_1\omega_2) \ , \tag{3}$$

where $n_\gamma(\omega)$ are the EPNs for a photon with energy ω and $\sigma^X_{\gamma\gamma}(\omega_1\omega_2)$ is the photon-photon cross section to produce of particle X. The photon-photon cross section can be calculated using Low's expression [54] based on the detailed balance theorem,

$$\sigma^X_{\gamma\gamma}(\omega_1\omega_2) = 8\pi^2(2J+1)\frac{\Gamma_{m_X\rightarrow\gamma\gamma}}{m_X}\, \delta\left(\omega_1\omega_2 - m_X^2\right) \tag{4}$$

Journal of Physics: Conference Series **2619** (2023) 012003 doi:10.1088/1742-6596/2619/1/012003

where J, m_X, and $\Gamma_{m_X \to \gamma\gamma}$ are the respective spin, mass and photon-photon $\gamma\gamma$ decay width of particle X. The delta-function enforces energy conservation. In Ref. [55] the relevance of different meson models and of exotic states is discussed thoroughly, including states that have not been considered before.

8.1. The Higgs

As an interlude, I'd like to mention that in 1989 a proposal emerged to search of the Higgs particle using UPCs at the LHC [56]. The production process is explained by means of Eqs. (3,4) with appropriate assumptions for the properties of the Higgs. We have made initial estimates in 1988 and obtained a cross section of 1 nb [31]. This is about the same value as the cross section for Higgs production at the LHC through hadronic processes with the advantage that the production of other particles is minimized. However, it was later noticed that another process, the direct photon-photon production of $b\bar{b}$ pairs, has a much larger cross section. Since the main mechanism of Higgs decay is through $b\bar{b}$ pairs, one concludes that the photon-photon production of the Higgs would be swamped in a background of directly produced pairs. The long sought Higgs particle was finally observed at the LHC in hadronic interactions [44, 45, 46].

8.2. Exotic mesons

Multiquark states such as multiquark molecules of the type $(q\bar{q})(q\bar{q})$, hybrid mesons $(q\bar{q}g)$ and glueballs (gg) are of large interest in meson spectroscopy [57]. UPCs can contribute to the search of multiquark resonances through anomalous $\gamma\gamma$ couplings and multiquark energy spectrum. UPCs might help to test the predictions of such "abnormal" states [44, 45, 46]. The $\gamma\gamma$ width is a measure of the charge of the constituent quarks. Therefore, the magnitude of the $\gamma\gamma$ coupling is useful to distinguish quark resonances from gluon-dominated resonances ("glueballs"). The absence of meson production via $\gamma\gamma$ fusion is also a useful signal for glueball search [44, 45, 46]. In UPCs a glueball can only be formed through annihilation of a $q\bar{q}$ pair into gluon pairs, whereas normal $q\bar{q}$ mesons are created directly.

8.3. Vector mesons and PDFs.

The production of vector mesons such as J/ψ and $\Upsilon(1s)$ can also be studied with Eq. (1) using the appropriate EPN. In 2001 [58] we were the first to propose this process to constrain generalized partonic distribution in nuclei, $F_A(x, Q^2)$, for a given momentum fraction x. For the real photon induced mechanism we used

$$\left.\frac{d\sigma^{\gamma A \to VA}}{dt}\right|_{t=0} = \frac{16\pi\alpha_s^2(Q^2)\Gamma_{ee}}{3\alpha M_V^5}\left[xF_A(x, Q^2)\right]^2, \tag{5}$$

where $\alpha_s(Q^2)$ is the strong interaction coupling evaluated at the perturbative Quantum Chromodynamics (pQCD) factorization scale $Q^2 = W_{\gamma g}^2$, M_V is the vector meson mass, Γ_{ee} is its leptonic decay width and $x = M_V^2/W_{\gamma p}^2$ is the fraction of the nucleon momentum carried by the gluons. The nuclear Parton Distribution Function (PDF), $F_a^A(\mathbf{r}, x, Q^2)$, can be written as a convolution of a medium modification function $R_a^A(\mathbf{r}, x, Q^2)$ with the nucleon PDF, $f_a(x, Q^2)$, with the subscript a denoting a parton species and the superscript A denoting a particular nucleus [59, 60], and $\mathbf{r}$ is the nucleon position within the nucleus. Nuclear modifications are included in $R_a^A(x, Q^2)$. For $x < 0.04$, one observes experimentally a shadowing effect, where the nuclear PDFs are smaller compared to the free nucleon distributions, $R_a^A < 1$ (Fig .9). In the region $0.04 < x < 0.3$ one observes an anti-shadowing effect, where $R_a^A > 1$. The EMC effect occurs in the range $0.3 < x < 0.8$. Another enhancement occurs for $x > 0.8$, due to the Fermi motion of nucleons. The physical principles governing these effects are quite different.

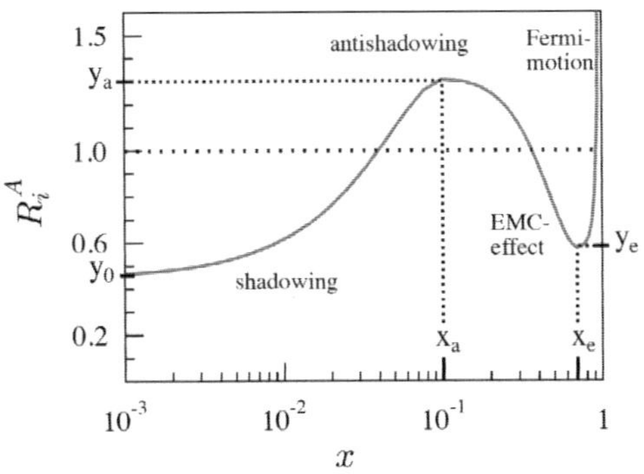

Figure 9. Medium modification function displaying numerous effects as a function of the momentum fraction x.

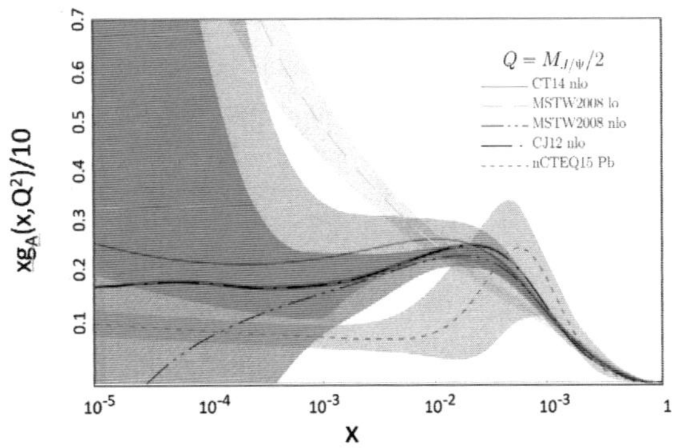

Figure 10. Uncertainties in theoretical compilations of gluon distribution functions (Courtesy of Fred Olness [61]).

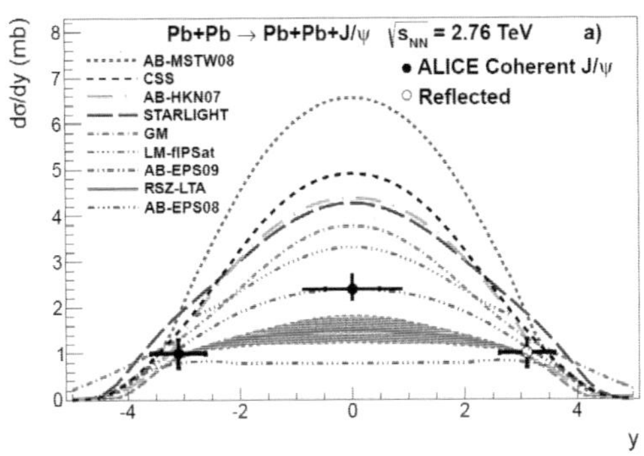

Figure 11. UPC production of J/Ψ at the LHC as a function of the rapidity y, compared to various PDFs (from Ref. [64]).

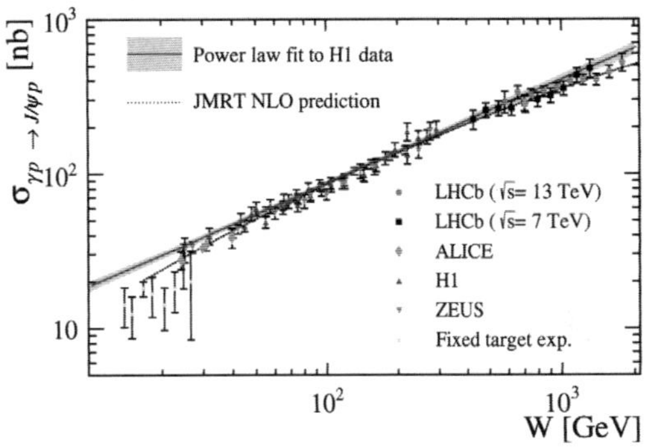

Figure 12. Compilation of cross sections for UPC production of J/Ψ in different nuclear facilities (from Ref. [65]).

In Refs. [59, 60] we have studied the impact of different gluon distributions in J/ψ and $\Upsilon(1s)$) production in UPCs. It is worthwhile noticing that UPCs with pPb and PbPb collisions lead to different production mechanisms called *direct and resolved.* The direct production implies that the photon interacts directly with the nucleus, while the resolved mechanism implies that the photon fluctuates into a quark-antiquark pair followed by a hadronic interaction with the nucleus. At leading order the direct production depends on the gluon distributions which are largely uncertain within the nucleus, specially at small x (Fig. 10 [61]). The resolved mechanism involves the distributions of light quarks and gluons in both photon and nucleus. The quadratic dependence of vector meson production in UPCs increases the sensitivity on gluon distributions of both cross sections and rapidity distributions [58, 59, 60].

Our calculations for J/Ψ production with gluon distributions that incorporate nuclear gluon shadowing [59, 60] is in good agreement with the ALICE experimental data [62, 63, 64], as shown in Fig. 11. It is now clear that J/Ψ and Υ photoproduction at the LHC are a powerful tool to study nuclear gluon shadowing in the region $x < 10^{-3}$. A compilation of cross sections at different facilities is shown in Fig. 12 showing a beautiful power law (see Ref. [65] for more details).

9. Physics beyond the Standard Model

In this brief review I have shown how genuinely simple ideas developed in the 1980s and 1990s have become a fruitful research area at the relativistic hadron colliders. Unexpected phenomena induced in UPCs have been verified experimentally such as double giant resonances, the dawn of studies with anti-atoms, light-by-light scattering, beam losses due to bound-electrons, probing parton distribution functions, and search for physics beyond the standard model. Could $\gamma\gamma \rightarrow$ graviton [67], multi-quark systems [66], or axion-like particles [32, 52, 53] be found in this way? The jury is out on whether this will be possible with further advances in new accelerators, beam physics and new detection techniques.

Acknowledgments

This work was partially supported by the U.S. DOE grant DE-FG02-08ER41533 and NSF grant no. 2114669 (Accelnet IANN-QCD https://www.iann-qcd.org).

References

[1] Bertulani C A and Baur G 1985 Relativistic Coulomb collisions *Nucl. Phys.* A 442 p 739.
[2] Fermi E 1924 On the Theory of Collisions between Atoms and Electrically Charged Particles *Zeit. Phys.* 29 p 315.
[3] Fermi E 1925 Sulla Teoria Dell' Urto Tra Atomi E Corpuscoli Elettici *Nuovo Cimento* 2 p 143.
[4] Weizsäcker C F Radiation due to collisions of very fast electrons *Z. Phys. 88* p 612.
[5] Williams E J 1934 Nature of the High Energy Particles of Penetrating Radiation and Status of Ionization and Radiation Formulae *Phys. Rev.* 45 p 724; Williams E J 1935 *K. Dan. Vidensk. Selsk. Mat.-Fys. Medd.* p 13 no. 4.
[6] Schwinger J S 1949 Quantum electrodynamics. III. The electrodynamic properties of the electron *Phys. Rev.* 76 p 790.
[7] Baur G 2008 Exchange of high and low energy photons in ultraperipheral relativistic heavy-ion collisions *Nucl. Phys. B* 179 p 129.
[8] Bertulani C A and Baur G 1988 Electromagnetic processes in relativistic heavy ion collisions *Phys. Reports* 163 p 299.
[9] Krauss F, Greiner M, Soff G 1997 Photon and gluon induced processes in relativistic heavy ion collisions *Prog. Part. Nucl. Phys.* 39 p 503.
[10] Baur G, Hencken K, Trautmann D, Sadovsky S and Kharlov Y 2002 Coherent $\gamma\gamma$ and γA interactions in very peripheral collisions at relativistic ion colliders *Phys. Reports* 364 p 359.
[11] Bertulani C A, Klein S R and Nystrand J 2005 Physics of ultra-peripheral nuclear collisions *Annu. Rev. Nucl. Part. Sci.* 55 p 271.
[12] Baur G, Hencken K and Trautmann D 2007 Electron-positron pair production in ultrarelativistic heavy ion collisions *Phys. Reports* 453 p 1.
[13] Baltz A J et al. 2008 The physics of ultraperipheral collisions at the LHC *Phys. Reports* 458 p 1.
[14] Baltz A J 2008 Evidence for higher order QED in e+ e-pair production at RHIC *Phys. Rev. Lett.* 100 062302.
[15] Nystrand J for the ALICE Collaboration 2008 Photon-Induced Physics with Heavy-Ion Beams in ALICE *Nucl. Phys. B (Proc. Suppl.)* 179 p 156.
[16] Baur G and Bertulani C A 1986 Electromagnetic processes in relativistic heavy ion collisions *Nucl. Phys. A* 458 p 725.
[17] Aumann T, Bertulani C A and Suemmerer K 1995 Neutron removal in peripheral heavy ion collisions *Phys. Rev. C* 51 p 416.
[18] Aumann T, Suemmerer K, Bertulani C A and Kratz J V 1996 Excitation of the two-phonon giant dipole resonance in 238U studied via inclusive measurements of neutron-removal cross sections *Nucl. Phys. A* 599 p 321.
[19] Aumann T, Borbtzky K, Stroth J, Wajda E, Blaich Th, Cub J, and Holeczek Jacek 1997 Coulomb fragmentation and Coulomb fission of relativistic heavy-ions and related nuclear structure aspects *Acta Physica Polonica. B* 1 p 375.
[20] Bertulani C A, Kucuk Y and Lozeva R 2020 Fission of relativistic nuclei with fragment excitation and reorientation *Phys. Rev. Lett.* 124 132301.
[21] Baltz A J, Chasman C and White S N 1998 Correlated forwardbackward dissociation and neutron spectra as a luminosity monitor in heavy-ion colliders *Nucl. Inst. Meth. Phys. Res. A* 41 p 1.
[22] Bertulani C A and Baur G 1994 (March) Relativistic heavy ion physics without nuclear contact *Physics Today* p. 22.

[23] Baur G and Bertulani C A 1986 Multiple electromagnetic excitations of giant dipole phonons in relativistic heavy ion collisions *Phys. Lett. B* 174 p 23.
[24] Baur G and Bertulani C A 1986 Multistep fragmentation of heavy ions in peripheral collisions at relativistic energies *Phys. Rev. C* 34 p 1654.
[25] Schmidt R et al. 1993 Electromagnetic excitation of the double giant dipole resonance in 136Xe *Phys. Rev. Lett.* 70 p 1767.
[26] Ritman J et al 1993 First observation of the Coulomb-excited double giant dipole resonance in 208 Pb via double-gamma decay *Phys. Rev. Lett.* 70 p 533.
[27] Emling 1994 Electromagnetic excitation of the two-phonon giant dipole resonance *Prog. Part. Nucl. Phys.* 33 p 729.
[28] Aumann T, Bortignon P F and Emling H 1998 Multiphonon giant resonances in nuclei *Ann. Rev. Nucl. Part. Sci.* 48 p 351.
[29] Bertulani C A and Ponomarev V 1999 Microscopic studies of the double giant resonance *Phys. Reports* 321 p 139.
[30] Baur G and Bertulani C A 1989 Electromagnetic processes in relativistic heavy ion collisions *Nucl. Phys. A* 505 p 835.
[31] Baur G and Bertulani C A, 1988 Electromagnetic physics at relativistic heavy ion colliders: for better and for worse *Z. Phys. A* 330 p 77.
[32] Adam W et al - CMS collaboration 2018 Evidence for light-by-light scattering and searches for axion-like particles in ultraperipheral PbPb collisions at $\sqrt{s}_{NN} = 5.02$ TeV *Phys. Lett. B* 797 p 303.
[33] Baur G and Bertulani C A 1987 Electromagnetic production of heavy leptons in relativistic heavy ion collisions *Phys. Rev. C* 35 p 836.
[34] Munger C and Brodsky S 1934 Production of relativistic antihydrogen atoms by pair production with positron capture *Physical Review D* 49 p 3228.
[35] Baur G et al 1996 Production of Antihydrogen *Physics Letters B* 368 p 3.
[36] Browne M W 1996 Physicists Manage to Create The First Antimatter Atoms *The New York Times* January 5.
[37] Blanford G et al 1998 Observation of Atomic Antihydrogen *Phys. Rev. Lett.* 80 p 3037.
[38] Bertulani C A and Baur G 1998 Theoretical calculation of antihydrogen production and accuracy of the equivalent photon approximation *Phys. Rev. D* 58 034005.
[39] Bertulani C A and Dolci D 2001 Pair production with capture in peripheral collisions with heavy ions *Nucl. Phys. A* 683 p 635.
[40] Klein S R 2014 Heavy ion beam loss mechanisms at an electron-ion collider *Phys. Rev.: Accelerators and Beams* 17 121003.
[41] Schaumann, Jowett J M, Bahamonde Castro C, Bruce R, Lechner A and Mertens T 2020 Bound-free pair production from nuclear collisions and the steady-state quench limit of the main dipole magnets of the CERN Large Hadron Collider *Phys. Rev.: Accelerators and Beams* 23 121003
[42] Reich E S 2010 Antimatter held for questioning *Nature News* 11 17.
[43] Grossman L 2010 The Coolest Antiprotons *Phys. Rev. Focus* 26 p 1.
[44] Bertulani C A and Navarra F 2002 Two- and three-photon fusion in relativistic heavy ion collisions *Nucl. Phys. A* 703 p 861.
[45] Moreira B D, Bertulani C A, Gonçalves V P and F. S. Navarra 2016 Production of exotic charmonium in gamma-gamma interactions at hadronic colliders *Phys. Rev. D* 94 094024.
[46] Bertulani C A, Gonçalves V P, Moreira B D and Navarra F S 2017 Production of exotic charmonium in ultra-peripheral heavy ion collisions *Eur. Phys. J.* 137 06019.
[47] Agakishiev H et al 2011 Observation of the antimatter helium-4 nucleus *Nature* 473 p 353.
[48] Bertulani C A and Ellermann M 2010 Production of exotic atoms at the CERN Large Hadron Collider *Phys. Rev. C* 81 044910.
[49] Baur G and Bertulani C A 1988 $\gamma\gamma$ physics with peripheral relativistic heavy ion collisions *Z. Phys. A* 330 p 77.
[50] Atlas collaboration 2017 Evidence for light-by-light scattering in heavy-ion collisions with the ATLAS detector at the LHC *Nature Physics* 13 p 852.
[51] Kusek-Gawenda M2019 Importance of mesons in light-by-light scattering in ultraperipheral lead-lead collisions at the LHC *EPJ Web Conf.* 199 05004.
[52] Gonçalves and Sauter W K 2020 Exclusive axionlike particle production by gluon - induced interactions in hadronic collisions, *Phys. Lett. B* 811 135981.
[53] Gonçalves V P, Martins D E and Rangel M S 2021 Searching for axionlike particles with low masses in pPb and PbPb collisions *Europ. Phys. J C* 81 p: 522.
[54] Low F E 1960 Proposal for Measuring the π^0 Lifetime by π^0 Production in Electron-Electron or Electron-

Positron Collisions *Phys. Rev.* 120 p 582.

[55] Bertulani C A 2009 Probing two-photon decay widths of mesons at energies available at the CERN Large Hadron Collider (LHC) *Phys. Rev. C* 79 047901.

[56] Papageorgiu E 1989 Coherent Higgs Production in Relativistic Heavy Ion Collisions *Phys. Rev. D* 40 p 92.

[57] Yao W-M et al 2006 Particle Data Group: Non-quark antiquark Mesons *J. Phys. G* 33 p 1.

[58] Gonçalves V P and Bertulani C A 2002 Peripheral heavy ion collisions as a probe of the nuclear gluon distribution *Phys. Rev. C* 65 054905. https://arxiv.org/abs/hep-ph/0110370

[59] Adeluyi A and Bertulani C A 2011 Gluon distributions in nuclei probed at the CERN Large Hadron Collider *Phys. Rev. C* 84 024916.

[60] Adeluyi A and Bertulani C A 2012 Constraining Gluon Shadowing Using Photoproduction in Ultraperipheral pA and AA Collisions *Phys. Rev. C* 85 044904.

[61] Olness F 2023 Private Communication.

[62] Abelev B et al 2013 (ALICE Collaboration) Coherent photoproduction in ultra-peripheral Pb-Pb collisions at $\sqrt{s_{NN}} = 2.76$ TeV *Phys. Lett. B* 718 p 1273.

[63] Abelev B et al 2013 (ALICE Collaboration) Charmonium and e^+e^- pair photoproduction at mid-rapidity in ultra-peripheral Pb-Pb collisions at $\sqrt{s_{NN}} = 2.76$ TeV *Eur. Phys. J. C* 73 p 2617.

[64] De Gruttola D (ALICE Collaboration) 2014 J/Ψ photoproduction in Pb-Pb and p-Pb ultra-peripheral collisions with ALICE at LHC *Nucl. Phys. A* 926 p 136.

[65] Alan et al - The LHCb collaboration 2018 Central exclusive production of J/Ψ and $\Psi(2S)$ mesons in pp collisions at $\sqrt{s} = 13$ TeV *J. High Energy Phys.* 10 p 167.

[66] Esposito A, Manzari C A, Pilloni A and Polosa A D 2021 Hunting for tetraquarks in ultraperipheral heavy ion collisions *Phys. Rev. D* 104 114029.

[67] Ahern S, Norbury J W and Poyser W J 2000 Graviton production in relativistic heavy-ion collisions *Phys. Rev. D* 62 116001.

Understanding Heavy-ion Fusion Cross Section Data Using Novel Artificial Intelligence Approaches

Daniele Dell'Aquila[1,2], Brunilde Gnoffo[3], Ivano Lombardo[3,4], Francesco Porto[3,5], Luigi Redigolo[3,4] and Marco Russo[3,4]

[1]Dipartimento di Fisica "Ettore Pancini", University of Naples "Federico II", Napoli, Italy
[2]INFN-Sezione di Napoli, Napoli, Italy
[3]Dipartimento di Fisica e Astronomia, University of Catania, Catania, Italy
[4]INFN-Sezione di Catania, Catania, Italy
[5]INFN-Laboratori Nazionali del Sud, Catania, Italy

E-mail: `daniele.dellaquila@unina.it`

Abstract. An unprecedentedly extensive dataset of complete fusion cross section data is modeled via a novel artificial intelligence approach. The analysis was focused on light-to-medium-mass nuclei, where fission-like phenomena are more difficult to occur. The method used to derive the models exploits a state-of-the-art hybridization of genetic programming and artificial neural networks and is capable to derive, in a data-driven way, an analytical expression that serves to predict integrated cross section values. We analyzed a comprehensive set of nuclear variables, including quantities related to the nuclear structure of projectile and target. In this paper, we describe the derivation of two computationally simple models that can satisfactorily describe, with a reduced number of variables and only a few parameters, a large variety of light-to-intermediate-mass collision systems in an energy domain ranging approximately from the Coulomb barrier to the oncet of multi-fragmentation phenomena. The underlying methods are of potential use for a broad domain of applications in the nuclear field.

1. Introduction

The cross section of the fusion between heavy ions has been deeply investigated in numerous experimental campaigns starting from approximately the 70s of the past century [1, 2, 3, 4, 5]. The underlying experimental techniques involve mainly gamma-ray analysis, time-of-flight and magnetic spectrometers, charged particle detection with telescope arrays. Data were collected especially above the Coulomb barrier and up to the onset of multi-fragmentation phenomena.

Our present understanding of the fusion cross section between heavy ions relies on a number of models, mainly semi-classical, which exploited large systematics of data [6]. In the case of light-to-medium-mass systems, i.e. systems in which the mass of the compound nucleus is not too large, $A_{tot} \simeq 20 - 140$, the fusion cross section at energies above the Coulomb barrier, as a function of the inverse center-of-mass energy $1/E_{cm}$, can be schematically divided into three main regions [2, 7]. The lowest energy region, usually called Region I, starts approximately at the Coulomb barrier and exhibits a quasi linear increase of the cross section for decreasing $1/E_{cm}$ values [1, 7, 2, 8, 9]. In this region, the fusion mechanisms is in competition with multi-nucleon-transfer processes [10, 11, 12] and quasi-fission (especially for heavy systems) [13, 14, 15]. At higher energies, in the so called Region II, the fusion cross section is typically characterized by a smooth fall. In this region, the competition with deep inelastic phenomena [16, 17] and quasi-fission makes it particularly challenging to correctly define the boundaries with region I. Correspondingly, reported data are often affected by relatively large uncertainties. After region II, the fusion cross section starts to decrease quite steeply with the energy. In this energy regime, referred as region III, fusion-evaporation is supplanted by incomplete fusion mechanisms [18, 19, 20], multi-fragmentation [21, 22, 23], and, in general, to much more complicated reaction scenarios [5]. At larger values of E_{cm}, the fusion cross section tends to negligibly small values, due to mechanical and thermodynamical instabilities of the transient system formed in the collisions [24, 25, 23, 22, 26, 27, 28].

The analysis of the fusion cross section at lower energies, in the sub-barrier domain, has seen an increase of popularity in the last decades [29]. On the other hands, the studies of heavy-ion fusion well above the Coulomb barrier (especially in regions II and III) were largely neglected after the 90s, leaving a number of unanswered questions that could be addressed with the help of modern detection systems for charged particles [30, 31, 32, 33, 34, 35, 36, 37, 38]. For example, one of the yet open questions regards the existence of a limiting angular momentum in the fusion between heavy ions, which was explicitly investigated only for a few systems [39]. The detailed interpretation of this portion of excitation functions, which is linked to the higher energy part, two main families of macroscopic models have been reported: critical distance models [7, 40] and models based on limitations to compound nucleus [41].

In addition to macroscopic models, also microscopical (Time-Dependent Hartree-Fock, TDHF, see e.g. [42, 43]), molecular dynamics (see e.g. [44, 45]), and phenomenological models (see, e.g., [1, 46, 47, 48, 28, 27]) have been proposed. The latter, in particular, achieved a noticeable precision in describing existing excitation functions of the fusion cross section Refs. [49, 28, 46, 48, 47] focusing, in particular, to the portion of the excitation function close to the maximum of the cross section. Recent phenomenological models, such as Ref. [27], are capable to describe the data in the highest energy region (III).

The phenomenological approaches mentioned above are useful to make numerical evaluations of the fusion cross section but have some limitations: (i) only a limited part of the energy regimes are usually described; (ii) they are often lacking physical boundaries.

Available models also leave unanswered questions regarding possible effects of nuclear structure parameters (i.e. characteristics of projectiles, targets and compound nuclei) on the fusion cross section. In this framework, we decided to analyze an unprecedentedly large dataset of complete fusion cross section data, with the aim to derive a universal model capable to describe the entire dataset and to inspect the correlation of the fusion cross section on nuclear structure variables.

This paper describes the derivation of new universal, data-driven models to describe the cross section of the fusion between light-to-medium-mass nuclei at energies above the Coulomb barrier. To derive the models described here, we exploited a novel artificial intelligence tool based on a hybridization of genetic programming and artificial neural networks [50], which provides state-of-the-art performance in analyzing even large datasets with numerous variables.

2. Dataset and Modeling

In this work, we fitted complete fusion data from 124 light-to-medium-mass collision systems ($Z_1 \cdot Z_2 \leq 250$) extracted from the dataset described in Refs. [51, 52]. The dataset used to derive the models contained about 4500 experimental points. Data associated with 63 medium-mass systems (with $Z_1 \cdot Z_2 > 250$) were excluded from the fit data, but were used to check the extrapolation of the models towards heavier collision systems.

In the present study, the fit was performed exploiting the Brain Project (BP) [50], a novel hybridization of genetic programming and artificial neural networks for the formal modeling of data. This tool was previously successfully used in a number of investigations in several research fields (see e.g. [53, 54, 55, 56]). BP belongs to the field of Evolutionary Computation [57], a widely used programming scheme that exploits concepts derived from the Darwinian theory of evolution to solve optimization problems (see also Ref. [58] for more details). In such a scheme, solutions of the problem to solve are suitably encoded to represent *individuals* of a *population*. A numerical value, called *fitness* and given by a fitness function, is then associated to each individual. The fitness function returns a value that represents the degree to which a given individual is a good solution for the specific problem to analyze. A series of iterations, which involve *crossover*, *mutation*, and other genetic operations are then performed with the result of improving the average fitness of the population. Iterations are stopped when a predefined criterion is met, and the individual with the best fitness value in the population is usually chosen as the best solution. More details regarding the implementation of BP can be found, for example, in Refs. [50, 59] and go beyond the scope of the present manuscript.

BP is capable to perform the so called *symbolic regression*. In a symbolic regression problem, some data are fitted according to a mathematical expression that is derived by the tool itself. In the present case, the target of the optimization problem is given by the fitness function, which contains the prediction error, i.e. to what extent the individual is a good fit of the data, and the mathematical complexity of the model. The latter is useful to find the correct trade-off between complexity and accuracy. For example, a too complex model would likely *overfit* the data, having poor generalization capabilities. On the other hand, a too simple model could provide insufficient level of accuracy. In the scheme of BP, the complexity is evaluated as the number of nodes (either variables, constants, or mathematical operations) required to compute the model. In the present application, we decided to use a given target number of nodes while deriving the models. In this way, BP tries to derive a model that exploits exactly the target number of nodes (n_{tgt}), while minimizing the prediction error.

According to a well defined scheme in machine learning applications, the dataset used to derive the models is suitably subdivided into learning and testing subsets. The learning dataset contained light-to-medium-mass systems, and was used to run BP and derive the models, while the testing dataset contained a random fraction of light-to-medium-mass systems and heavier systems, as elucidated above. The various collision systems are represented in Fig. 1 in the Z_1 vs Z_2 plane. In this representation, the horizontal axis always identifies the charge of the lighter collision partner, for clarity reasons. In the figure, collision systems belonging to the learning dataset are shown in black, while collision systems in the first testing dataset (the one used to test the extrapolations towards heavier systems) are shown in blue, and the second testing set is represented by red squares. The latter comprises experimental data from all systems present in the learning set plus additional low-statistics systems not represented in the learning dataset. The size of each box is such that the square of its area is proportional to the statistics in in the corresponding Z_1 vs Z_2 combination.

A key point of this study is the advanced *feature selection*, i.e. the capability of an artificial intelligence tool to automatically select the input variables with the largest importance in predicting the target output (i.e. the cross section of the fusion between heavy ions, in this application), neglecting other poorly important variables. This is made possible by the mechanism implemented by BP [54]. In this work, we exploited the advanced feature selection capabilities of BP to explore the dependence on several variables, including nuclear structure variables that identify projectile, target, and compound systems, of the fusion cross section between heavy ions. Table 1 shows a list of the variables used in this study.

3. Results

To derive the models discussed in this paper, we run BP with number of target nodes n_{tgt} ranging between 10 and 20, as discussed in details in Ref. [60]. In this way, we obtain gradually more complex models, up to a maximum of $n_{tgt} = 20$. We observe that the prediction accuracy

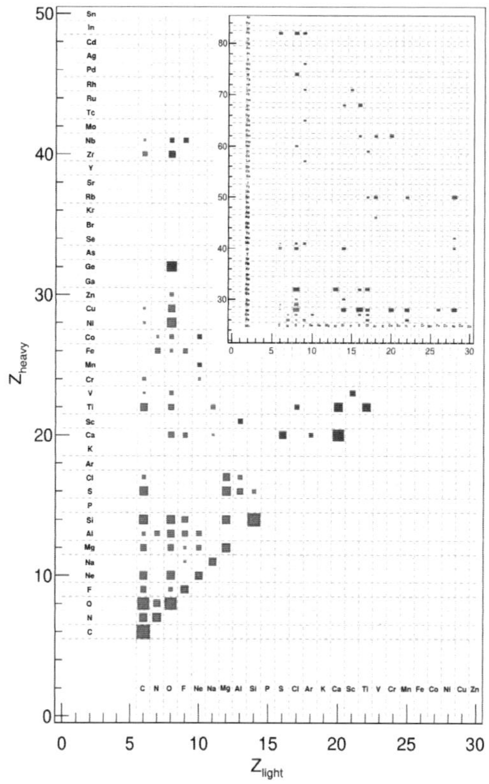

Figure 1. A schematic representation of the dataset used to derive the models in the Z_1 vs Z_2 plane. The charge of the heaviest fragment in the collision system is represented in the x-axis. Points in the learning dataset are represented with gray boxes, points in the testing dataset are represented with red boxes. Blue boxes are testing systems with $Z_1 \cdot Z_2 > 250$. The square of the area of each box is proportional to the statistics in in the corresponding Z_1 vs Z_2 combination.

increases up to $n_{tgt} \approx 15$, and then it completely saturates. Driven by these preliminary checks, we derived two different models, using $n_{tgt} = 10$ and $n_{tgt} = 15$, which have the following mathematical expressions:

$$\sigma_{fus}^{(n_{tgt}=10)}(E_{cm}) = 1103 \cdot \exp\left[-\left(1.387 - 0.468 \cdot \frac{Z_2 \cdot Z_1}{E_{cm}}\right)^2\right] \tag{1}$$

$$\sigma_{fus}^{(n_{tgt}=15)}(E_{cm}) = 1116 \cdot \exp\left[-\sinh^2\left(-1.359 + \mathrm{erf}\left(\frac{S_{2n}}{E_{cm}}\right) + 0.061 \cdot \frac{A_1 \cdot A_2}{E_{cm}}\right)\right] \tag{2}$$

where both formulas are expressed in units of mb. The first formula is extremely simple and handful. It is very convenient to be used for fast evaluations of the fusion cross section of a given collision system at a given energy. This simplest model exploits only 3 features: E_{cm}, Z_1 and Z_2. Interestingly, it foresees the product $Z_1 \cdot Z_2$, which is related to the Coulomb term. Several other phenomenological models show a clear dependence on $Z_1 \cdot Z_2$ (e.g., [49, 28, 47, 48]). The complex model derived in this work, $n_{tgt} = 15$, exploits 4 features: E_{cm}, A_1, A_2, and S_{2n}. In this case, we find the product $A_1 \cdot A_2$, which has a similar dependence than $Z_1 \cdot Z_2$ found in the simplest equation. Due to the weak dependence on the two-neutron separation energy of the

Symbol	Description
E_{cm}	collision center-of-mass energy (MeV)
Z_1	charge of the first reaction partner
Z_2	charge of the second reaction partner
N_1	number of neutrons of the first reaction partner
N_2	number of neutrons of the second reaction partner
A_1	mass of the first reaction partner
A_2	mass of the second reaction partner
J_1	spin of the first reaction partner
J_2	spin of the second reaction partner
π_1	parity of the first reaction partner (1 for positive parity, −1 for negative parity)
π_2	parity of the second reaction partner (1 for positive parity, −1 for negative parity)
μ_1	magnetic dipole momentum of the first reaction partner (μ_N)
μ_2	magnetic dipole momentum of the second reaction partner (μ_N)
$\langle r^2 \rangle_1$	rms charge radius of the first reaction partner (fm)
$\langle r^2 \rangle_2$	rms charge radius of the second reaction partner (fm)
Q-value	fusion Q-value (MeV)
S_α	α separation energy of the compound nucleus (MeV)
S_n	one-neutron separation energy of the compound nucleus (MeV)
S_{2n}	two-neutron separation energy of the compound nucleus (MeV)
S_p	one-proton separation energy of the compound nucleus (MeV)
S_{2p}	two-proton separation energy of the compound nucleus (MeV)
$S_{\alpha 1}$	α separation energy of the first reaction partner
$S_{\alpha 2}$	α separation energy of the second reaction partner
S_{n1}	one-neutron separation energy of the first reaction partner
S_{n2}	one-neutron separation energy of the second reaction partner
S_{p_1}	one-proton separation energy of the first reaction partner
S_{p_2}	one-proton separation energy of the second reaction partner
S_{2n1}	two-neutron separation energy of the first reaction partner
S_{2n2}	two-neutron separation energy of the second reaction partner
S_{2p_2}	two-proton separation energy of the first reaction partner
S_{2p_2}	two-proton separation energy of the second reaction partner

Table 1. List of variables (features) used to derive the models discussed in this paper.

compound system, in the complex model, for neutron-rich systems, one finds a slight increase of the complete fusion cross section at energies larger than the optimal fusion energy. This finding is in qualitative agreement with both the phenomenological model of Ref. [49] and with experiments at the onset of multi-fragmentation [24, 61]. No particularly significant correlation on other features related to nuclear structure properties can be found with this approach. This facts suggests that the informative content of present nuclear fusion datasets might be essentially contained in the set of variables E_{cm} and Z_1Z_2 (or A_1A_2 and S_{2n}).

Figure 2 shows the results of our $n_{tgt} = 10$ (red dashed line) and $n_{tgt} = 20$ (red solid line) models for some typical collision systems in our dataset. For comparison, we report also the results of Ref. [49] (blue dashed line). In the figure, open circles represent experimental points of light-to-medium-mass systems not considered in the learning dataset, while open triangles are used to indicate data from heavy systems used to test the extrapolation capabilities of our models. From the figure, it is evident that the agreement with the data is improved compared to previous state-of-the-art phenomenological models, especially in the regions close

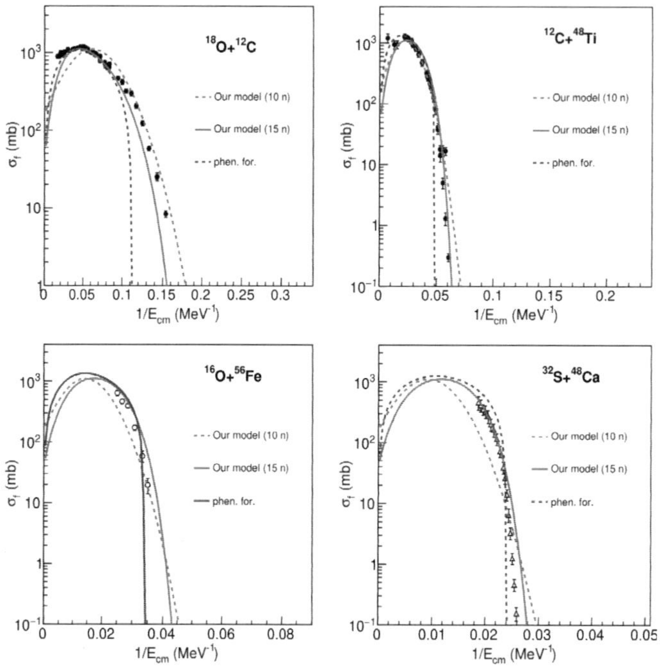

Figure 2. Results of our models against experimental data. Open symbols represent data excluded from the learning dataset. Blue lines represent the results of Ref. [49].

to the Coulomb barrier, where, for example, the model of Ref. [49] falls to rapidly to zero and even to negative values of the cross section. In our models, the cross section at low energy has an exponential-like fall that closely follow experimental data points. The results are also satisfactory for experimental data not included in the learning dataset (see open symbols).

4. Conclusion and Perspectives

In the present work, we have fitted an unprecedentedly dataset of fusion cross section data between light-to-medium-mass nuclei above the Coulomb barrier using state-of-the-art artificial intelligence tools for symbolic regression. We exploited a novel hybridization of genetic programming and artificial neural networks to produce mathematical expressions capable to fit the data.

We propose a computationally simple model and a more complex one, with slightly improved performance. We find the further complicating the models does not improve the accuracy. In both cases, a few variables, mainly E_{cm}, Z_1, Z_2 (or A_1, A_2, and S_{2n}) are found to be sufficient to describe the entire systematics. The inclusion of additional variables is found not to add further informative content.

This study has a double impact: (1) it provides computationally simple models to evaluate the cross section of the fusion between heavy ions; (2) it helps identifying the variables correlated with the fusion cross section, exploring also an extensive set of nuclear structure parameter. More recently, this study has supported a novel investigation, see Ref. [62], in which the role of shell effects in the fusion between heavy ions is discussed in details.

References

[1] Glas D and Mosel U 1975 *Nucl. Phys. A* **237** 429–440

[2] Bass R 1980 *Nuclear Reactions with heavy Ions* 1st ed (Sunderland, MA: Springer-Verlag) ISBN 978-3-540-09611-5
[3] Sanders S J, Toledo A S d and Beck C 1999 *Phys. Rep.* **311** 487–551
[4] Back B B *et al.* 2014 *Rev. Mod. Phys.* **86**(1) 317–360
[5] Jha V, Parkar V V and Kailas S 2020 *Phys. Rep.* **845** 1–58
[6] Fröbrich P 1984 *Phys. Rep.* **116** 337–400
[7] Lee S M, Matsuse T and Arima A 1980 *Phys. Rev. Lett.* **45** 165–168
[8] Pakou A *et al.* 2015 *Eur. Phys. J. A* **51** 55
[9] Mazzocco M *et al.* 2015 *Phys. Rev. C* **92** 024615
[10] Corradi L, Pollarolo G and Szilner S 2009 *J. Phys. G.: Nucl. Part. Phys.* **36** 113101
[11] von Oertzen W and Vitturi A 2001 *Rep. Prog. Phys.* **64** 1247–1337
[12] Mijatović T 2022 *Frontiers in Physics* **10** 965198
[13] Tõke J, Bock R, Dai G X, Gobbi A, Gralla S, Hildenbrand K D, Kuzminski J, Müller W F J, Olmi A, Stelzer H, Back B B and Bjørnholm S 1985 *Nucl. Phys. A* **440** 327–365
[14] Hinde D J, Jeung D Y, Prasad E, Wakhle A, Dasgupta M, Evers M, Luong D H, du Rietz R, Simenel C, Simpson E C and Williams E 2018 *Phys. Rev. C* **97** 024616
[15] Kozulin E M *et al.* 2020 *Eur. Phys. J. A* **56** 6
[16] Moretto L G and Schmitt R P 1981 *Rep. Prog. Phys.* **44** 533–591
[17] Beck C, Djerroud B, Haas F, Freeman R M, Hachem A, Heusch B, Morsad A, Youlal M, Abe Y, Dayras A, Wieleczko J P, Matsuse B T and Lee S M 1992 *Zeitschrift fur Physik A Hadrons and Nuclei* **343** 309–323
[18] Hanold K, Moretto L G, Peaslee G F, Wozniak G J, Bowman D R, Mohar M F and Morrissey D J 1993 *Phys. Rev. C* **48** 723–734
[19] Gerschel C 1982 *Nucl. Phys. A* **387** 297–312
[20] Samarin V V 2018 *Phys. Atom. Nuc.* **81** 486–496
[21] Bougault R *et al.* 2018 *Phys. Rev. C* **97** 024612
[22] Borderie B *et al.* 2018 *Phys. Lett. B* **782** 291–296
[23] De Filippo E *et al.* 2009 *Acta Phys. Pol. B* **40** 1199
[24] Amorini F *et al.* 2009 *Phys. Rev. Lett.* **102** 112701
[25] Cardella G *et al.* 2012 *Phys. Rev. C* **85** 064609
[26] Manduci L *et al.* 2016 *Phys. Rev. C* **94** 044611
[27] Eudes P *et al.* 2014 *Phys. Rev. C* **90** 034609
[28] Giordano R, Sambataro S, Scalia A, Porto F, P, Figuera and Pirrone S 1990 *Nuov. Cim. A* **103** 47
[29] Jiang C L, Back B B, Rehm K E, Hagino K, Montagnoli G and Stefanini A M 2021 *Eur. Phys. J. A* **57** 235
[30] Dell'Aquila D *et al.* 2018 *Nucl. Instr. Meth. Phys. Res. A* **877** 227
[31] Acosta L *et al.* 2016 *J. Phys.: Conf. Ser.* **730** 012001
[32] Bishop J *et al.* 2019 *Phys. Rev. C* **100** 034320
[33] Cardella G *et al.* 2015 *Nucl. Instr. Meth. Phys. Res. A* **799** 64
[34] Dell'Aquila D *et al.* 2016 *Nuov. Cim. C* **39** 272
[35] Lopez O *et al.* 2018 *Nucl. Instr. Meth. Phys. Res. A* **884** 140
[36] Pastore G *et al.* 2017 *Nucl. Instr. Meth. Phys. Res. A* **860** 42
[37] Camaiani A *et al.* 2021 *Phys. Rev. C* **103** 014605
[38] Dell'Aquila D *et al.* 2019 *Nucl. Instr. Meth. Phys. Res. A* **929** 162
[39] Beck C and Szanto de Toledo A 1996 *Phys. Rev. C* **53** 1989–1992
[40] Matsuse T, Arima A and Lee S M 1982 *Phys. Rev. C* **26** 2338–2341
[41] Bass R 1977 *Phys. Rev. Lett.* **39** 265–268
[42] Reinhard P G *et al.* 2016 *Phys. Rev. C* **93** 044618
[43] Zheng H *et al.* 2018 *Phys. Rev. C* **98** 024622
[44] Maruyama T *et al.* 2002 *J. Nucl. Rad. Sci.* **3** 77
[45] Feng Z Q, Jin G M and Zhang F S 2008 *Nucl. Phys. A* **802** 91–106
[46] Horn D and Ferguson A J 1978 *Phys. Rev. Lett.* **41** 1529–1532
[47] Lozano M and Madurga G 1980 *Phys. Lett. B* **90** 50
[48] Kailas S and Gupta S K 1981 *Z. Phys. A* **302** 355
[49] Porto F and Sambataro S 1984 *Nuov. Cim.* **83** 339
[50] Russo M 2016 *Swarm Evo. Comput.* **27** 145
[51] Karpov A V *et al.* 2017 *Nucl. Instrum. Meth. Phys. Res. A* **859** 112
[52] Zagrebaev V I, Denikin A S, Karpov A V, Alekseev A P, Naumenko M A, Rachkov V A, Samarin V V and Saiko V V 1999 NRV web knowledge base on low-energy nuclear physics URL `http://nrv.jinr.ru/`
[53] Russo M, Leotta G, Pugliatti P M and Gigliucci G 2014 *Solar Energy* **105** 264–273
[54] Campobello G, Dell'Aquila D, Russo M and Segreto A 2020 *Appl. Soft Comput.* **94** 106488

[55] Buccheri E, Dell'Aquila D and Russo M 2021 *Diabetes Res. Clin. Pract.* **174** 108722
[56] Buccheri E, Dell'Aquila D and Russo M 2022 *Obes. Med.* **31** 100398
[57] Koza J R 1992 *Genetic Programming: On the Programming of Computers by Natural Selection* (Cambridge, MA, USA: MIT Press)
[58] Dell'Aquila D and Russo M 2021 *Comput. Phys. Commun.* **259** 107667
[59] Russo M 2020 *Soft Comput.* **24** 16885
[60] Dell'Aquila D, Gnoffo B, Lombardo I, Porto F and Russo M 2023 *Jour. Phys. G.: Nucl. Part. Phys.* **50** 015101
[61] D'Agostino M *et al.* 2011 *Nucl. Phys. A* **861** 47–66
[62] Dell'Aquila D, Gnoffo B, Lombardo I and Porto F 2023 *Phys. Lett. B* **837** 137642

Dynamics in Nuclei

Ani Aprahamian[1,2], **Shelly R. Lesher**[3], **Kevin Lee**[1]

[1] Department of Physics & Astronomy, University of Notre Dame, Notre Dame, IN 46556, USA

[2] A.I. Alikhanyan National Science Laboratory, 2 Alikhanyan Bros. Street, 0036 Yerevan, Armenia

[3] Physics Department, University of Wisconsin-LaCrosse, LaCrosse, WI, USA

E-mail: aapraham@nd.edu

Abstract. The nucleus is a complex many-body system with some remarkable emergent collective properties of multiple nucleons acting together. Bohr and Mottelson [1] provided a description of collective motion in nuclei based on geometrical shapes with superimposed oscillations around those shapes. Later, Lie algebras and symmetries were used to describe nuclear dynamics [2] , followed by advances in the shell model approach [3] with new effective nucleon-nucleon two- and three-body interactions, and more recently with Hartree-Fock-Bogoliubov approximations within the extended generator coordinate method [4]. Yet, the underlying science question has remained the same. In nuclei, where there is explicit deformation in the ground state, "are the low-lying 0^+ states collective vibrations built on the ground state or are they minima of a coexisting shape?" Ref. [4] has shown that for a significant percentage of $K = 0^+$ excitations built on the deformed ground state (g.s.) should, in fact, be a collective vibration. The question has remained open due to sufficiently convincing experimental data with lifetimes, transfer reaction cross sections, and $E0$ transitions [5]. This paper summarizes the experimental situation regarding the lifetimes of 0^+ states.

1. Introduction

The nucleus is a complex many-body system with some remarkable emergent properties of multiple nucleons acting together to result in what can be described as superimposed oscillations of the nucleus around that shape. This resulted in the 1975 Physics Nobel Prize to Bohr, Mottelson, and Rainwater [1]. Later, Lie algebras and symmetries were highly successful in describing the observed dynamics within group theory representations [2] describing limits of nuclear collective behavior and resulting in predictions that explained the connection of the low-lying excited bands in deformed nuclei. Hence, in the ensuing four decades, many other theoretical and experimental developments have been made including the implementation of new effective nucleon-nucleon two- and three-body interactions within the shell model [3], and more recently with Hartree-Fock-Bogoliubov approximations within the extended generator coordinate method [4]. Yet, the underlying science question has remained the same. In nuclei, where there is explicit deformation in the ground state, "are the 0^+ states collective vibrations built on the ground state or are they minima of a coexisting shape?" The shapes that occur in the low-energy part of the spectra of nuclei can be described by a quadrupole ($\lambda = 2$) deformation in nature, resulting in two types of vibrations in deformed nuclei: β and γ oscillations, the latter breaking axial symmetry with a symmetry axis projection of $K^\pi = 2^+$. The γ vibration is

Journal of Physics: Conference Series **2619** (2023) 012005 doi:10.1088/1742-6596/2619/1/012005

characterized as the first $K^\pi = 2_1^+$, or 2_γ^+ band. The transition probabilities across the deformed rare-earth region of nuclei is typically in the order of a few Weisskopf units $[B(E2 : 2_\gamma^+ \to 0_{g.s.}^+)]$ (W.u.) [6].

The story however is different for the K=0^+ excitations [4]. A significant percentage of them are collective vibrations and built on the deformed g.s. The absence of collective excitations in nuclei would uniquely distinguish nuclei in the quantum world of atoms, molecules, etc.

This question has remained mostly unresolved in nuclear physics due to the lack of sufficiently comprehensive experimental data with lifetimes, transfer reaction cross-sections, and $E0$ transitions [5].

Experiments carried out at the high precision Q3D spectrometer of the University of Munich have revolutionized nuclear structure studies by showing that there are tens of low-lying excited $K^\pi = 0^+$ states [7, 8] below the pairing gap in well-deformed nuclei. Figure 1 summarizes the results of the various Q3D experiments. There are several 0^+ states observed in the rare-earth nuclei (Sm, Gd, Dy, Er, Yb, and Hf). In the ^{154}Gd nucleus, there are sixteen 0^+ states identified, in the ^{162}Dy nucleus, there are twelve 0^+ states reported [8], while in the ^{168}Er case, there are fifteen 0^+ states identified. The origin and characterization of these states remain an open challenge, and it is the observation of so many $K^\pi = 0^+$ states that have revitalized the discussion about the viability of vibrational excitations.

Following the remarkable Q3D measurements, we have measured the lifetimes of the observed 0^+ states where possible. From the results of our work, we present examples of double-phonon $\gamma\gamma$, $\beta\beta$, and a $\beta\gamma$ vibration.

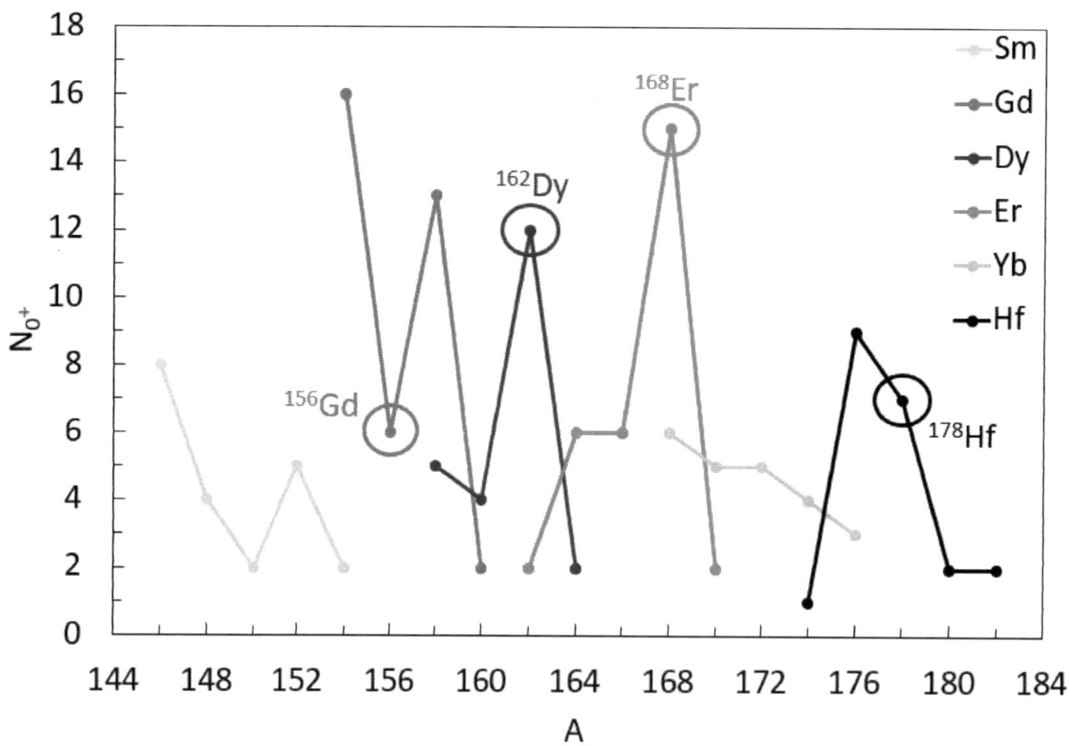

Figure 1. (Color online) The confirmed and tentative 0^+ states observed in several rare-earth nuclei (Sm, Gd, Dy, Er, Yb, & Hf).

2. Experiments

The lifetime measurements reported here include the GRID technique [9, 10]. The technique involves the measurement of γ-line widths following neutron capture. Experiments have been carried out at the Institute Laue-Langevin (ILL) neutron High Flux Reactor in Grenoble, France. They also include results from the DSAM technique [11, 12, 13] which involves the $(n, n'\gamma)$ reaction at the University of Kentucky Accelerator Laboratory (UKAL).

2.1. GRID

The GRID technique was used to measure the level lifetimes in ^{162}Dy and ^{156}Gd [6, 14] at the High Flux Reactor at the Institut Laue-Langevin (ILL) in Grenoble, France. The GRID technique [10, 15] is based on precisely measuring the width of a decaying gamma-ray using perfect crystals. The initial recoiling velocity of the nucleus results in a broadening of the width of a given gamma-ray emitted in flight. Knowing the slowing down process from simulations, the nuclear level lifetimes are extracted. The recoiling velocities are typically in the 10^{-4}c to 10^{-6}c range which results in a few eV broadening of that γ-ray. This precise measurements of the energy or wavelength of the γ rays are possible with Si or Ge perfect crystal diffraction.

2.2. $n, n'\gamma$

Level lifetimes for ^{162}Dy [16] and ^{168}Er [17] nuclei have been measured with inelastic neutron scattering at the University of Kentucky Accelerator Laboratory (UKAL). Measurements include γ-ray excitation functions and angular distributions. The emitted γ rays are usually measured with a $\leq$50% efficient HPGe detector suppressed by a BGO shield [11].

3. Results

3.1. ^{156}Gd

The rare-earth deformed region includes the Gd isotopes. The $R_{4^+/2^+}$ ratio is 3.24 in the ^{156}Gd nucleus. The spectra in the low-lying part of the excitation spectra of this nucleus is well known up to an excitation energy of 2.35 MeV. In this range of energy, there are six excited $K^\pi = 0^+$ bands, where four of them are below the 2 MeV pairing gap. The low-lying level scheme of ^{156}Gd is shown in fig. 2.

Two of the $K^\pi = 0^+$ bands in this nucleus are connected to the ground state band. Transitions from the one $K^\pi = 0^+_2$ band at 1049.5 keV are more collective than the ones from the other $K^\pi = 0^+_3$ band at 1168.2 keV. Plotting the moments of inertia for the various $K^\pi = 0^+$, the $K^\pi = 2^+$, and the $K^\pi = 4^+$ bands are very informative. It appears that all the bands except the $K^\pi = 0^+_3$ band at 1168.2 keV have nearly identical moments of inertia with the ground state band. This is evidence of collectivity built on the ground state for all of the bands except the $K^\pi = 0^+_3$ band [17]. The transition probabilities calculated for transitions from the $K^\pi = 2^+$ band to the ground state band also support the assignment of this individual band as the γ band. The $K^\pi = 0^+_4$ and the $K^\pi = 4^+_1$ bands that are at 1715.1 keV and 1510.6 keV, respectively, show strong connections to the $K^\pi = 2^+$ γ band. This is evidence of a second set of two-phonon $\gamma\gamma$ vibrational excitations although the anharmonicities are very different when compared with the data for the ^{166}Er nucleus.

3.2. ^{162}Dy

Similarly, our new measurements of lifetimes in the ^{162}Dy nucleus allow us to say that the K^π=0^+ excitation at 1400.2 keV is, in fact, the β vibrational excitation. The earlier assumption that this band is perhaps the two-phonon $\gamma\gamma$ vibration is disputed. The new measurements point to significant collectivity above the pairing gaps. Specifically, a 4^+ state at 2180.6 keV with $B(E2)$=8.4 W.u. to the K^π=2^+ γ band showing fragmentation of the K^π=4^+ strength to

Journal of Physics: Conference Series **2619** (2023) 012005 doi:10.1088/1742-6596/2619/1/012005

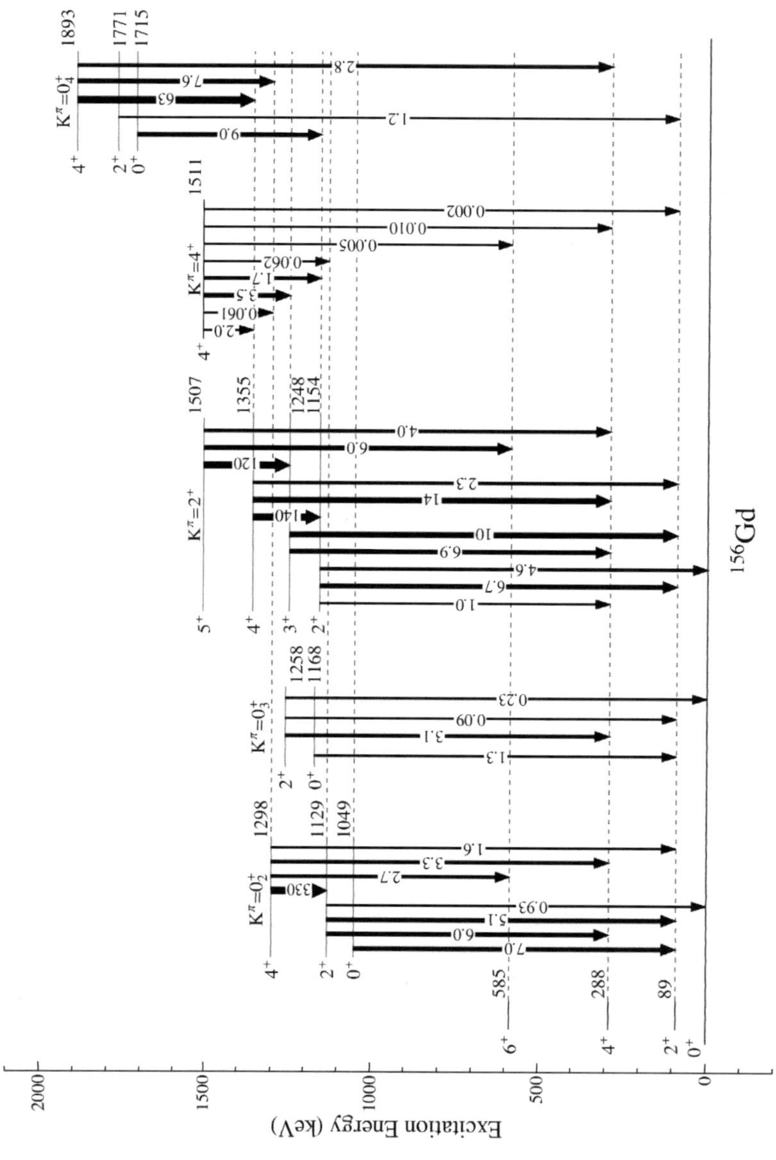

Figure 2. The ^{156}Gd level scheme with levels with new lifetimes measurements. The GRID values are displayed with the use of a ratio of $0.6 \times \tau_{\text{max}}$ of the GRID range. This is described in the text. The level with $K^\pi = 4^+$, $J^\pi = 4^+$ was measured earlier by coulomb excitation. We display widths of transitions in proportion the transition probability values in W.U.

Table 1. Ratios of B($E2$) values and Energies in comparison to theoretical values. There are two values if fragmentation is observed and measured.

Ratios	^{156}Gd	^{162}Dy	^{168}Er	Theory
$E(4^+_{\gamma\gamma})/E(2^+_\gamma)$	1.31	1.73; 2.5	2.5	2.0
$E(0^+_{\gamma\gamma})/E(2^+_\gamma)$	1.49	1.6		2.0
$E(2^+_{\beta\gamma})/E(2^+_\gamma)$		2.5		≈ 2.0
$B(E2{:}4^+_{\gamma\gamma} \to 2^+_\gamma)/B(E2{:}2^+_\gamma \to 0^+_{g.s})$	0.39	14±9; 1.8		2.78
$B(E2{:}0^+_{\gamma\gamma} \to 2^+_\gamma)/B(E2{:}2^+_\gamma \to 0^+_{g.s})$	1.96			5.0
$B(E2{:}2^+_{\beta\gamma} \to 2^+_\gamma)/B(E2{:}2^+_\gamma \to 0^+_{g.s})$		1.1		1.43

the K^π=2^+ band. This is similar to the case in ^{166}Er [18]. Also, a state at 2231.0 keV has a B($E2$)=5.1 W.u transition to the same K^π=2^+ band, indicating that it may be the first example of a two phonon $\beta\gamma$ vibration. Fig. 3 shows the part of the level scheme of interest.

3.3. ^{168}*Er*

The lifetimes and the associated B($E2$) values are shown in Fig. 4. In this nucleus, we report on the collectivity of the third excited 0^+ state. There is once again, a strong similarity between the ^{166}Er case and the ^{168}Er nucleus. It is once again, the fourth 0^+_4 state that carries the collectivity. The theoretical predictions from Ref. [4] indicate that the collectivity to the ground state should be at the at 1.818 MeV which corresponds to the third excited K^+=0^+ state [17].

4. Discussion

The lifetimes of various deformed nuclei have been measured and interesting characteristics emerge for the 0^+ bands. In the ^{156}Gd, there are two $K^\pi = 0^+$ bands low in the excitation spectrum of the nucleus. The 0^+ band at 1049.5 keV is more collective than the $K^\pi = 0^+$ band at 1715.1 keV. This latter band is connected to the γ band or the first excited $K^\pi = 2^+$ band. In the ^{162}Dy nucleus, we find the fragmented strength of two-phonon $\gamma\gamma$ vibration and for the first time, a two-phonon $\beta\gamma$ vibration. In the ^{168}Er nucleus, there are several 0^+ states the one with the greater transition probabilities connecting to the ground state is the third excited $K^\pi = 0^+$ band. We present the information on single and double-phonon excitations in Table 1 in comparison with theory.

In this paper, we show an extensive set of lifetime measurements that complement transfer reaction data. In order to develop a comprehensive understanding of collectivity across this region of deformation, additional measurements of $E0$ transitions are required. Only then, do we hope to develop a consistent picture that can help us differentiate between collectivity and coexistence in nuclei.

5. Acknowledgments

This research was supported by the US NSF under contracts PHY-2011890 and PHY-2011267.

References

[1] Bohr A and Mottelson B 1975 *Nuclear Structure Vol. II* (Reading, Massachusetts: W.A. Benjamin)
[2] Iachello F and Arima A 1987 *The Interacting Boson Model* (Cambridge University Press)
[3] Caurier E, Martínez-Pinedo G, Nowacki F, Poves A and Zuker A P 2005 *Rev. Mod. Phys.* **77**(2) 427–488
[4] Delaroche J P, Girod M, Libert J, Goutte H, Hilaire S, Péru S, Pillet N and Bertsch G F 2010 *Phys. Rev. C* **81** 014303
[5] Heyde K and Wood J L 2011 *Rev. Mod. Phys* **83** 1467
[6] Aprahamian A, Lesher S R, Casarella C, Börner H G and Jentschel M 2017 *Phys. Rev. C* **95** 024329

[7] Lesher S R, Aprahamian A, Trache L, Oros-Peusquens A, Deyliz S, Gollwitzer A, Hertenberger R, Valnion B D and Graw G 2002 *Phys. Rev. C* **66** 051305(R)
[8] Meyer D A, Wood V, Casten R F, Fitzpatrick C R, Graw G, Bucurescu D, Jolie J, von Brentano P, Hertenberger R, Wirth H F, Braun N, Faestermann T, Heinze S, Jerke J L, Krücken R, Mahgoub M, Möller O, Mücher D and Scholl C 2006 *Phys. Rev. C* **74** 044309
[9] Börner H G, Jolie J, Hoyler F, Robinson S, Dewey M S, Greene G, Kessler E and Deslattes R D 1988 *Phys. Lett. B* **215** 45
[10] Börner H 1988 *I.O.P. Conf. Ser.* **88** 143
[11] Garrett P E, Warr N and Yates S W 2000 *J. Res. Natl. Inst. Stand. Technol.* **105** 141
[12] Shelton E and Rodgers V C 1973 *Comput. Phys. Commun.* **6** 99
[13] Belgya T, Molnár G and Yates S W 1996 *Nucl. Phys. A* **607** 43
[14] Aprahamian A, de Haan R C, Lesher S R, Casarella C, Stratman A, Börner H G, Lehmann H, Jentschel M and Bruce A M 2018 *Phys. Rev. C* **98** 034303
[15] Jolie J, Ulbig S, Börner H G, Lieb K P, Robinson S J, Schillebeeckx P, Kessler E G, Dewey M S and Greene G L 1989 *Europhysics Letters* **10** 231
[16] Aprahamian A, Lesher S R, Casarella C, Lee K, Crider B P, Lowe M, Meier M M, Peters E E, Prados-Estévez F M, Ross T J, Tully Z and Yates S W 2023 *Phys. Rev. C* **submitted** x
[17] Lesher S, Aprahamian A, Lee K, Alemayehu B, Clark L, James X, Lowrie J, Meier M, McEwan L, Mukhopadhyay S, Peters E E, Ramirez A, Ryan M, Rice B, Stratman A, Temanson E, Vanhoy J and Yates S 2022 *Phys. Rev. C* **106** 044302
[18] Garrett P E, Kulp W D, Wood J L, Bandyopadhyay D, Choudry S, Dashdorj D, Lesher S R, McEllistrem M T, Mynk M, Orce J N and Yates S W 2009 *Phys. Rev. Lett.* **103** 062501
[19] Aprahamian A, Wu X, Lesher S R, Warner D D, Gelletly W, Börner H G, Hoyler F, Schreckenback K, Casten R F, Shi Z R, Kusnezov D, Ibrahim M, Macchiavelli A O, Brinkman M A and Becker J A 2006 *Nucl. Phys. A* **764** 42
[20] Bucurescu D, Graw G, Hertenberger R, Wirth H, Iudice N L, Sushkov A V, Shirikova N Y, Sun Y, Faestermann T, Krücken R, Mahgoub M, Jolie J, von Brentano P, Braun N, Heinze S, Möller O, Mücher D, Scholl C, Casten R F and Meyer D A 2006 *Phys. Rev. C* **73** 064309
[21] Mang H, Poggenburg J and Rasmussen J 1965 *Nuclear Physics* **64** 353–378 ISSN 0029-5582
[22] Wang M, Huang W, Kondev F, Audi G and Naimi S 2021 *Chinese Physics C* **45** 030003

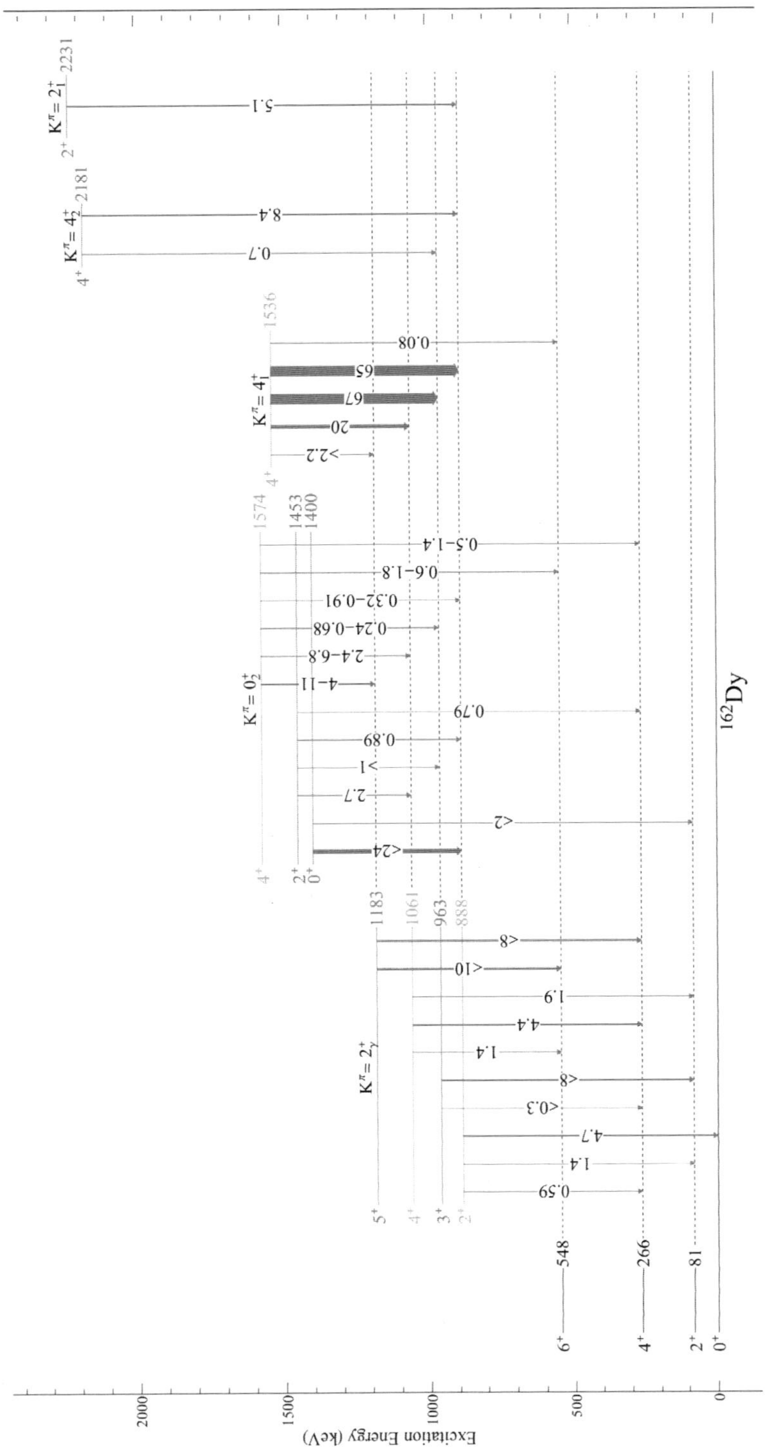

Figure 3. (Color online) Level scheme showing all observed transitions from the lowest-lying 2_γ^+, 0_2^+, and 4_1^+ band with all interband ($K^\pi = J_i^+ \rightarrow K^\pi = 2_\gamma^+$) decays. Levels with new lifetime measurements are drawn in red while those with an associated literature lifetime are drawn in turquoise (color online). The B($E2$) calculations are expressed in Weisskopf units (W.u.) and drawn in red. Intensities for transitions to the γ band are taken from [19]. Also shown is a tentative (2^+) state at 2231.0 keV with a strong decay to the 2_γ^+ band.

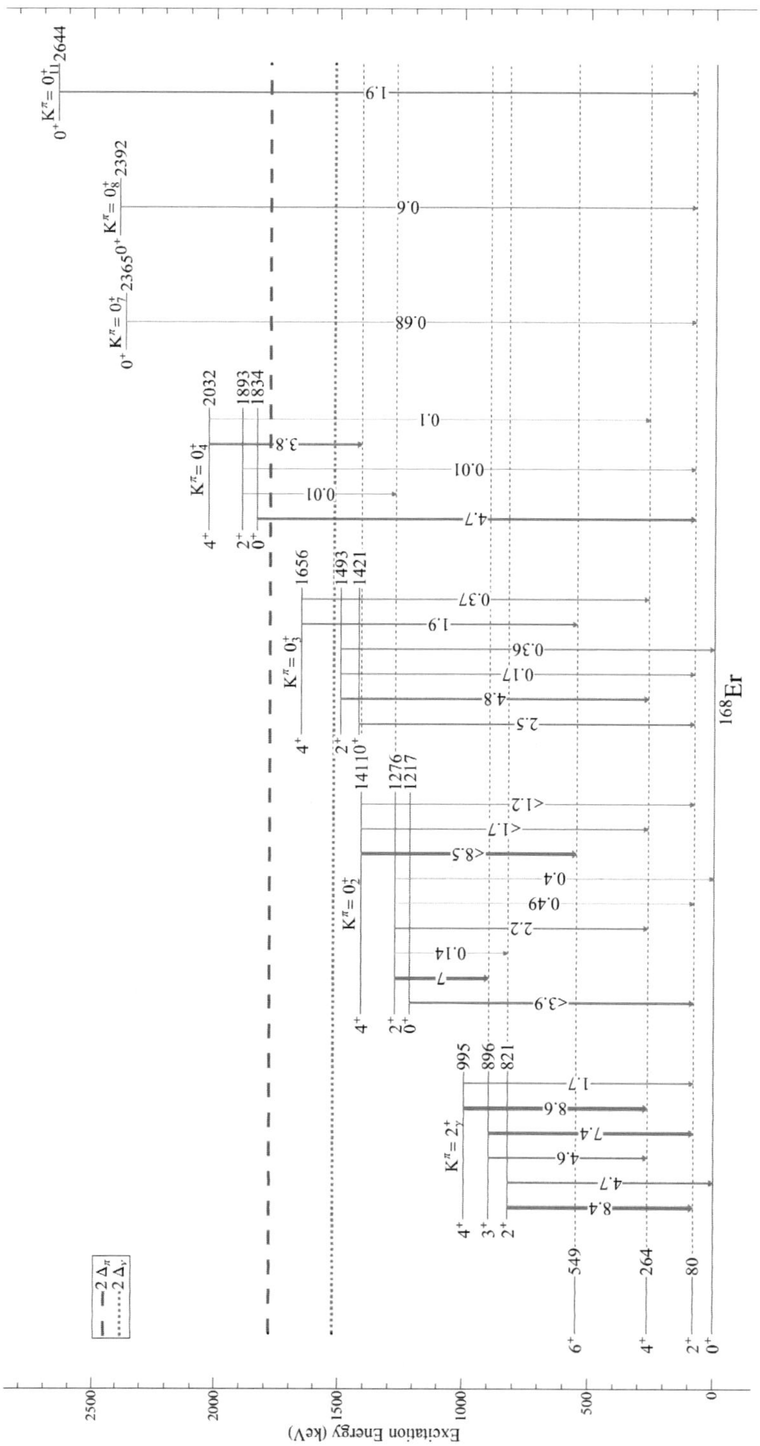

Figure 4. (Color online) Partial level scheme for the 0^+ bands identified and reported in Ref. [17] and discussed in this work. The solid arrows represent $E2$ transitions. The band labels are according to the (p,t) experiment labels [20]. The dotted horizontal lines are the two-neutron and two proton pairing gaps for ^{168}Er calculated using the formulas in Ref. [21] and masses from Ref. [22].

^{10}Be in the Cluster Shell Model

Omar Alejandro Díaz Caballero

Instituto de Ciencias Nucleares, UNAM, AP 70-543, Coyoacán, 04510 Ciudad de México, México

E-mail: odiazcab@gmail.com

Roelof Bijker

Instituto de Ciencias Nucleares, UNAM, AP 70-543, Coyoacán, 04510 Ciudad de México, México

E-mail: bijker@nucleares.unam.mx

Abstract. The Cluster Shell Model describes light nuclei in terms of clusters of k α-particles and x extra nucleons, in which the extra nucleons move in the deformed field generated by geometric configuration of α-particles. We present the first study for the case of $x = 2$ nucleons in an application to ^{10}Be as a cluster of two α-particles and two neutrons.

1. Introduction

The study of cluster structures in light nuclei goes all the way back to the pioneering contributions from Wheeler [1] and Hafstad and Teller [2] in the 1930s and Brink in the 1960s [3]. A recent review can be found in Ref. [4].

In the year 2000 an algebraic method, called the Algebraic Cluster Model (ACM), was proposed to describe light nuclei as clusters of α-particles [5]. As an example, the nucleus ^{12}C was described successfully in the ACM as a triangular configuration of three α-particles [6], in particular the $L^P = 5^-$ state was predicted more than a decade before its experimental discovery [7].

In subsequent years, the Cluster Shell Model (CSM) was introduced to describe the properties of neigboring cluster nuclei of the type $k\alpha + x$ as composed of k α-particles plus $x = 1$ extra nucleon [8, 9, 10, 11]. The aim of this contribution is to present the first application for $x = 2$ extra nucleons. As an example, we study the nucleus ^{10}Be as a cluster of $k = 2$ α-particles and $x = 2$ neutrons, and investigate to what extent the cluster structure of two α-particles persists under the addition of two neutrons.

2. Cluster Shell Model

The Cluster Shell Model describes configurations of light nuclei of the type $k\alpha+x$, *i.e.* composed of k α-particles plus x extra nucleons, in which the extra nucleons move in the deformed field generated by the cluster of α-particles [8]. The case of one extra nucleon with $x = 1$ was studied for ^{9}Be and ^{9}B ($k = 2$) [9], ^{13}C ($k = 3$) [10], and ^{21}Ne and ^{21}Na ($k = 5$) [11].

In this contribution, we present the first application for two extra nucleons for the case of the nucleus ^{10}Be as a cluster of $k = 2$ α-particles and $x = 2$ neutrons. The neutron single-particle levels are described by the CSM Hamiltonian

$$H_{\rm CSM} = T + V(\vec{r}) + V_{so,r} + \frac{1}{2}(1+\tau_3)V_C(\vec{r}) \ , \tag{1}$$

which is the sum of the kinetic energy, a central potential, a spin-orbit interaction and a Coulomb potential for the case of an odd proton. The central potential is obtained by convoluting the density

$$\rho(\vec{r}) = \left(\frac{\alpha}{\pi}\right)^{\frac{3}{2}} \sum_{i=1}^{2} e^{-\alpha(\vec{r}-\vec{r_i})^2} \ . \tag{2}$$

with the nucleon-α interaction [?]. The coefficient α is related to the size of the α-particle, and the $\vec{r_i}$'s represent the coordinates of the α-particles: $\vec{r}_1 = (\beta, 0, -)$ and $\vec{r}_2 = (\beta, \pi, -)$ where β denotes the distance of the α-particles with respect to the center of mass.

The CSM wave functions are calculated in the intrinsic or body-fixed frame. A system of two identical α-particles has axial symmetry, and hence the eigenstates are characterized by K^P: the projection of the angular momentum along the symmetry axis K and the parity P. The calculations are carried out in the harmonic oscillator basis. The i-th eigenstate is given by the expansion

$$\mid K_i^P \rangle = \sum_{nlj} C_{nljK}(i) \mid n, (l, \tfrac{1}{2}), j, K\rangle \, \delta_{P,(-1)^l} \ . \tag{3}$$

Figure 1 shows the splitting of the spherical single-particle levels in the deformed field generated by the cluster of two α-particles as a function of β [8, 9]. For $\beta = 0$ we recover the spherical harmonic oscillator limit, whereas for $\beta > 0$ the harmonic oscillator levels are mixed and split into levels characterized by K^P. For the case of interest, the value of β is determined from the moment of inertia of the rotational band in ^{8}Be as $\beta = 1.82$ fm [9]. The five lowest eigenstates are listed in Table 1.

3. The nucleus ^{10}Be

In this section, we discuss the nucleus ^{10}Be as a cluster of $k = 2$ α-particles and $x = 2$ neutrons. In addition to the CSM Hamiltonian of Eq. (1) we have to add the residual interaction between the two neutrons. Here we propose to use a pairing interaction [12]

$$\begin{aligned} V_{\rm pair} &= -G \sum_{\mu,\nu>0} a^\dagger_\mu a^\dagger_{\bar{\mu}} a_{\bar{\nu}} a_\nu \\ &= -G \sum_{\mu,\nu>0} Q_+(\mu) Q_-(\nu) \ . \end{aligned} \tag{4}$$

Table 1. Eigenvalues of CSM in MeV

	K_i^P	Energy
$1s_{\frac{1}{2}}$	$\frac{1}{2}^+_1$	-10.38
$1p_{\frac{3}{2}}$	$\frac{1}{2}^-_1$	-3.81
$1p_{\frac{3}{2}}$	$\frac{3}{2}^-_1$	-1.75
$1p_{\frac{1}{2}}$	$\frac{1}{2}^-_2$	0.35
$1d_{\frac{5}{2}}$	$\frac{1}{2}^+_2$	1.38

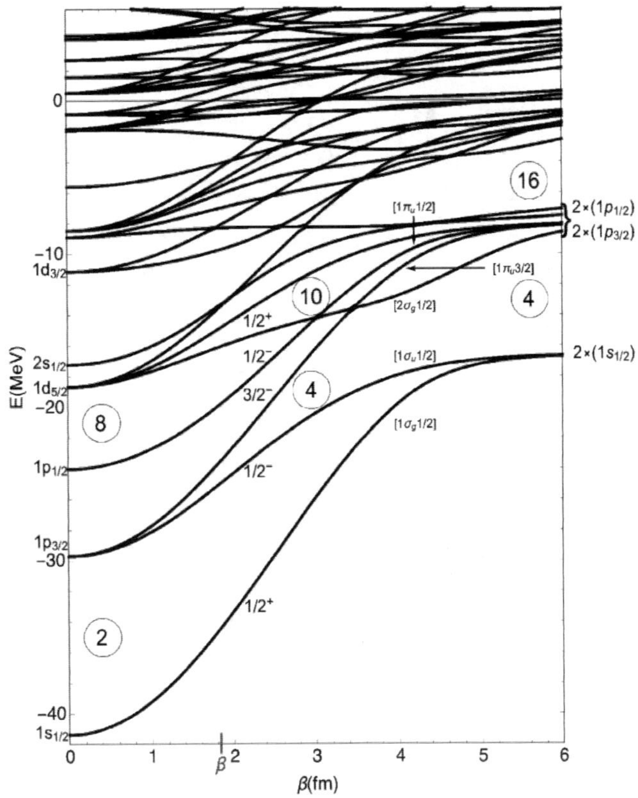

Figure 1. Excited states of the ^{9}Be as a function of β

In this notation $\bar{\mu}$ represents the time-reversed partner of μ which is degenerate with μ. The pair creation and annihilation operators, $Q_+(\mu)$ and $Q_-(\nu)$, together with the number operator

$$Q_0(\mu) = \frac{1}{2}\left(a^\dagger_\mu a_\mu - a_{\bar{\mu}} a^\dagger_{\bar{\mu}}\right) , \tag{5}$$

satisfy the $SU(2)$ fermion quasi-spin algebra

$$\begin{aligned} [Q_+(\mu), Q_-(\nu)] &= 2\delta_{\mu\nu} Q_0(\mu) , \\ [Q_0(\mu), Q_\pm(\nu)] &= \pm\delta_{\mu\nu} Q_\pm(\mu) . \end{aligned} \tag{6}$$

The two-particle states are given by

$$\mid \mu\rangle = Q_+(\mu) \mid 0\rangle , \qquad \langle \mu \mid = \langle 0 \mid Q_-(\mu) . \tag{7}$$

The matrix elements of the pairing interaction in the two-paricle states are

$$\begin{aligned} \langle \mu \mid V_{pair} \mid \nu\rangle &= -G \sum_{\mu'\nu'} \langle 0 \mid Q_-(\mu) Q_+(\mu') Q_-(\nu') Q_+(\nu) \mid 0\rangle \\ &= -G . \end{aligned} \tag{8}$$

The relevant single-particle levels are shown in Table 1. The first two levels with $K_i^P = \frac{1}{2}^+_1$ and $\frac{1}{2}^-_1$ are occupied by the four neutrons of the two α-particles. In our calculation, we consider the next three available levels with $K_i^P = \frac{3}{2}^-_1$, $\frac{1}{2}^-_2$ and $\frac{1}{2}^+_2$ for the two extra neutrons.

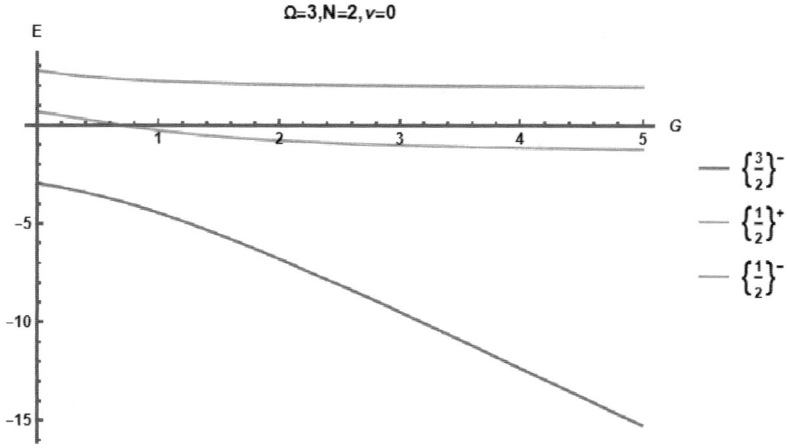

Figure 2. $K^P = 0^+$ energies as a function of the pairing strength, G.

The only configurations affected by the pairing interaction are the $K^P = 0^+$ states. The eigenvalues can be obtained by diagonalizing the CSM Hamiltonian plus the pairing interaction, $H_{CSM} + V_{\text{pair}}$, with matrix elements

$$\begin{pmatrix} -G + 2E_3 & -G & -G \\ -G & -G + 2E_4 & -G \\ -G & -G & -G + 2E_5 \end{pmatrix} \tag{9}$$

Figure 2 shows the behavior of the $K^P = 0^+$ eigenstates as a function of the pairing strength, G. The value of the pairing strength is determined from the excitation energy of the first excited 0^+ state to be $G =$??? MeV. The final results for ^{10}Be are shown in Figure 3.

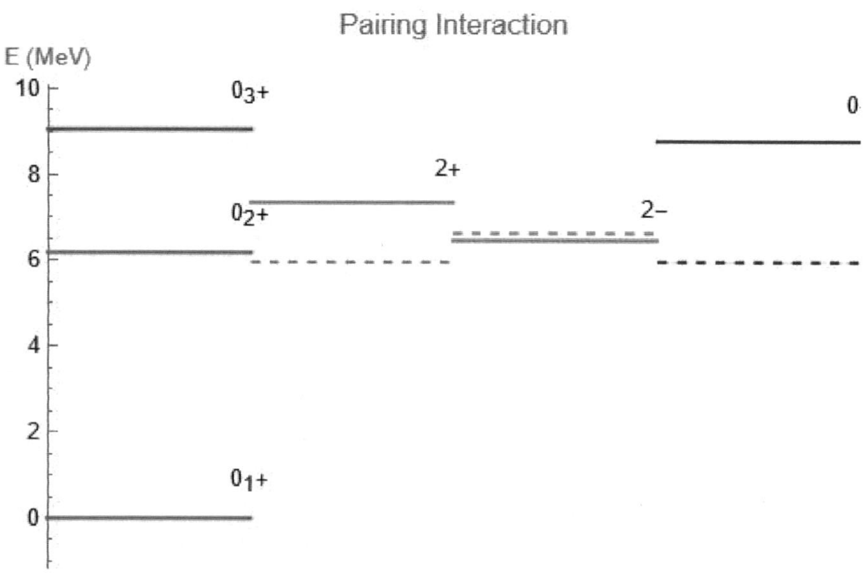

Figure 3. Paring interaction and the experimental data (dotted line)

4. Summary and conclusions

We presented the first extension of the Cluster Shell Model to a cluster of α-particles with two extra nucleons with an application to the nucleus ^{10}Be. The residual interaction between the two extra neutrons was taken as a pairing potential whose strength was adjusted to the excitation energy of the first excited 0^+ state. We find a good agreement with the available experimental data. A more detailed CSM analysis of ^{10}Be including other forms of the residual interaction as well as a study of transition probabilities will be published elsewhere [13].

Acknowledgments

This work was supported in part by grant No. IG101423 from PAPIIT, DGAPA-UNAM, Mexico.

References

[1] J.A. Wheeler, Phys. Rev. **52** (1937) 1083.
[2] L.R. Hafstad and E. Teller, Phys. Rev. **54** (1938) 681.
[3] D.M. Brink, in: Proc. Int. School of Physics "Enrico Fermi", Course XXXVI (1965) p. 247.
[4] M. Freer, H. Horiuchi, Y. Kanada-En'yo, D. Lee and U.-G. Meissner, Rev. Mod. Phys. **90** (2018) 035004.
[5] R. Bijker and F. Iachello, Phys. Rev. C **61** (2000) 067305.
[6] R. Bijker and F. Iachello, Ann. Phys. (N.Y.) **298** (2002) 334.
[7] D.J. Marín-Lámbarri, R. Bijker, M. Freer, M. Gai, Tz. Kokalova, D.J Parker and C. Wheldon, Phys. Rev. Lett. **113** (2014) 012502.
[8] V. Della Rocca, R. Bijker and F. Iachello, Nucl. Phys. A **966** (2017) 158.
[9] V. Della Rocca and F. Iachello, Nucl. Phys. A **973** (2018) 1.
[10] R. Bijker and F. Iachello, Phys. Rev. Lett. **122** (2019) 162501.
[11] R. Bijker and F. Iachello, Nucl. Phys. A **1010** (2021) 122193.
[12] O. Burglin and N. Rowley, Nucl. Phys. A **602** (1996) 21.
[13] O.A. Díaz-Caballero and R. Bijker, in preparation.

Journal of Physics: Conference Series **2619** (2023) 012007 doi:10.1088/1742-6596/2619/1/012007

Nuclear structure and dynamics with light exotic beams

A Di Pietro[1]
[1]INFN, Laboratori Nazionali del Sud, via S. Sofia 62, 1-95123 Catania, Italy
E-mail: dipietro@lns.infn.it

Abstract. In this contribution some aspects of the reaction dynamics of collisions involving light exotic nuclei will be introduced. In particular the dynamics of proton-halo nuclei and the differences with neutron-halo induced reactions will be discussed. Moreover, the formation of quasi-molecular states in the scattering of light nuclei having a neutron skin with α-particles will be presented.

1. Introduction

Unstable nuclei in the low mass region of the nuclear chart have yielded many surprising structures, among others the halo structure for nuclei in the neutron and proton drip-line. In the case of light nuclei, both stable and radioactive, other types of structures may appear, like molecular-like structures where α particles are bound by the exchange of ns or the existence of cluster configurations where at least one of the clusters is a soft nucleus. Using post-accelerated radioactive ion beams there is the opportunity to study nuclear structure and reactions of these exotic nuclei.

In this contribution an overview of some of the new phenomena involving light exotic RIBs will be given. In particular, the results of a reaction induced by the proton-halo nucleus ^{8}B on a ^{64}Zn target will be reported and compared with reactions induced by neutron-halo on the same target. Many efforts have been made over the years to understand the dynamics of reactions at the barrier with halo nuclei. The dynamics of proton-halo nuclei is expected to be different than neutron-halo. ^{8}B, with its very low breakup threshold of 0.138 MeV, is a good candidate for having a proton-halo structure. All ^{8}B experiments, previous to the one reported here, have been done using in-flight ^{8}B beam, therefore with some limitations associated with the characteristics of the beams (purity, energy spread and beam emittance) compared to the post-accelerated ones.

In the elastic scattering process of neutron-halo nuclei, coupling to the continuum results in a suppression in the elastic scattering angular distribution relative to the Rutherford [e.g. [1, 2, 3, 4, 5] ensuing a large total reaction cross-section. Moreover, a large fraction of this cross-section is due to direct processes such as break-up or transfer [2, 6, 7]. Experiments with ^{8}B beam, at energies around the Coulomb barrier [8, 9], found a total reaction cross-section as large as in the case of neutron-halo reactions. In the case of ^{8}B breakup high order effects had to be considered to reproduce the data. The availability at ISOLDE of a post accelerated ^{8}B beam gives the possibility to investigate the above topic with a better precision than the one so far available.

This manuscript reports also on a different aspect of the interplay between nuclear structure and nuclear reaction in the case of light-neutron-rich nuclei. In particular, the investigation of the elastic scattering process of the neutron-rich ^{9}Li nucleus on a ^{4}He target. In [10], the calculations reported show that the reaction ^{9}Li+^{4}He, in the entrance channel, is affected by the compound nucleus ^{13}B rotational band structure. The mass density plots in [10], show that molecular systems can be formed between the ^{9}Li and ^{4}He, hence rotational bands can be excited in the ^{9}Li+^{4}He reaction. For some given impact parameter, the α is captured by the extended neutron tail of ^{9}Li which exerts a weak attraction. The system begins to rotate until the ^{4}He will be ejected back into the elastic channel or produces other reaction channels via exciting inelastic degrees of freedom. As a consequence the elastic excitation function exhibits broad energy resonances. In addition, elastic cross-section may be enhanced at backward angles due to the orbiting of the ^{4}He about the ^{9}Li ion.

To understand the nature of such broad structures an orbiting reaction process was investigated which gave a reasonable explanation of the results.

2. The ^{8}B+^{64}Zn experiment

The experiment was performed at HIE-ISOLDE facility at CERN with the first post-accelerated ^{8}B beam [11]. The ^{8}B beam was produced from the protons coming from the CERN PSBooster with energy of 1.4 GeV through the reaction on a a multiwalled carbon nanotube target (CNT). A molecular ^{8}BF$_2$ beam was extracted from the target; ^{8}B was charge bred in REX-EBIS to 3^{+} and accelerated to an energy of 4.9 A MeV. The final ^{8}B average intensity on target was $\sim$400pps. The reaction target was a 1.021 mg/cm^2 isotopically enriched ^{64}Zn target. This was placed at an angle of 30° with respect to the beam direction, in order to allow measurements at 90°. The experimental setup consisted of six silicon telescopes made of two stages of detectors, 40 μm and 1000 μm thick respectively. Each detector was a 50x50 mm^2 Doube Sided Silicon Strip detector (DSSSD) segmented into 16+16 strips [12]. The angular distribution was measured with high resolution, limited only by the statistics.

The ^{8}B+^{64}Zn elastic scattering angular distribution is shown in fig. 1 as a ratio to the Rutherford cross-section. A clear Coulomb-nuclear interference peak is visible at around 35°, differing from the neutron-halo nucleus ^{11}Be on the same target and at similar $E_{\mathrm{c.m.}}/V_{\mathrm{C}}$ [4, 5].

The measured angular distribution is compared with continuum-discretized coupled-channels (CDCC) calculations. In the calculations ^{8}B is described as a two-cluster system, a ^{7}Be and a proton. Details of the calculations are reported in [11]. Figure 1 shows the result of the calculations.

The results of the CDCC calculations are shown in fig. 1 as a solid line. The calculation omitting the coupling to the continuum channels (dashed line) is also shown for comparison. From the comparison of the calculations with and without continuum coupling, as well as from the experimental results, it is possible to observe that the elastic breakup produces a small suppression of the elastic cross-section, and does not suppress completely the Coulomb-nuclear interference effect. The total reaction cross-section from CDCC calculations is σ_{R} =1.5 b, smaller than the one obtained with the n-halo ^{11}Be (σ_{R}=2.7 b) on the same target and at similar $E_{\mathrm{c.m.}}/V_{\mathrm{C}}$. This result appears to be at variance with what it was reported in [13]. Figure 3 shows the comparison of the ^{11}Be+^{64}Zn [4, 5] and ^{8}B+^{64}Zn elastic scattering angular distributions at similar $E_{\mathrm{c.m.}}/V_{\mathrm{C}}$. It can be seen that the angular distribution in the case of ^{8}B differs from that of the n-halo ^{11}Be, mainly in the Coulomb-nuclear interference region. This is an indication that ^{8}B does not show the strong effects of long-range absorption of the n-halo case.

In the Coulomb break-up process, proton-halo nuclei are expected to behave differently from neutron-halo, since the loosely bound proton takes part to the reaction. Dynamic polarization effects displace the valence proton behind the nuclear core shielding it from the target causing

Journal of Physics: Conference Series **2619** (2023) 012007 doi:10.1088/1742-6596/2619/1/012007

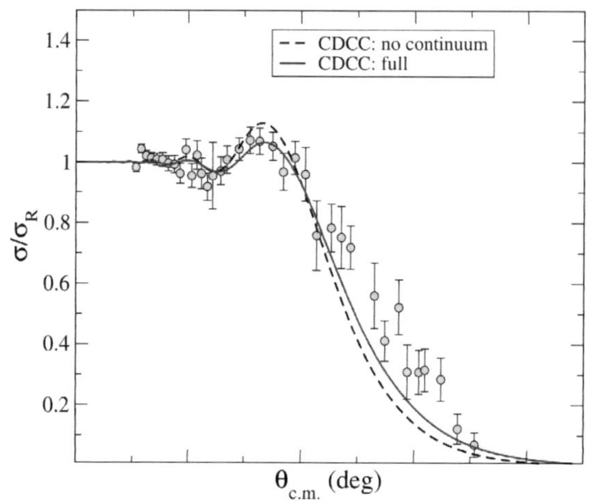

Figure 1. Experimental elastic scattering angular distribution for $^8B+^{64}Zn$ (symbols). CDCC calculations with (full line) and without (dashed line) coupling to the continuum.

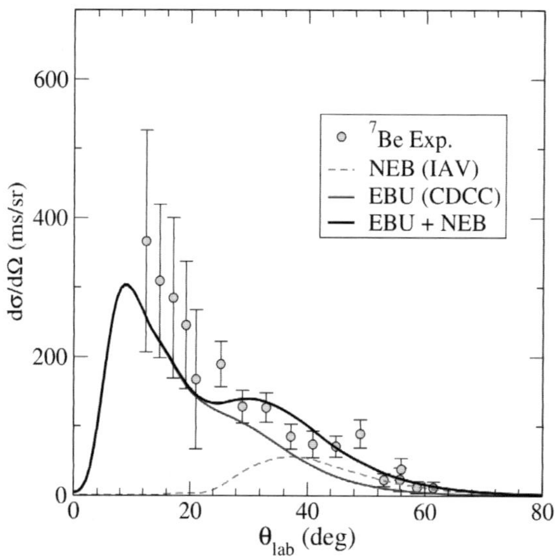

Figure 2. Experimental ^{7}Be angular distribution (symbols). Continuous blue line: EBU from CDCC calculations. Dashed red line: NEB. Continuous black line: sum of EBU+NEB contributions.

a reduction of the break-up probability compared to first-order perturbation theory predictions and higher-order corrections are required [e.g. [14, 15]]. Figure 2 shows the angular distribution for ^{7}Be events coming from the ^{8}B breakup. The present breakup data is inclusive, i.e. several reaction processes can in principle contribute to the ^{7}Be yield: the elastic breakup (EBU) process, in which the ^{8}B is dissociated into ^{7}Be+p and the target is in the ground state; the non-elastic breakup (NEB) processes in which the dissociated proton interacts with the target. This last process includes: non-capture breakup accompanied by target excitation, proton absorption by the target (incomplete fusion) and proton transfer.

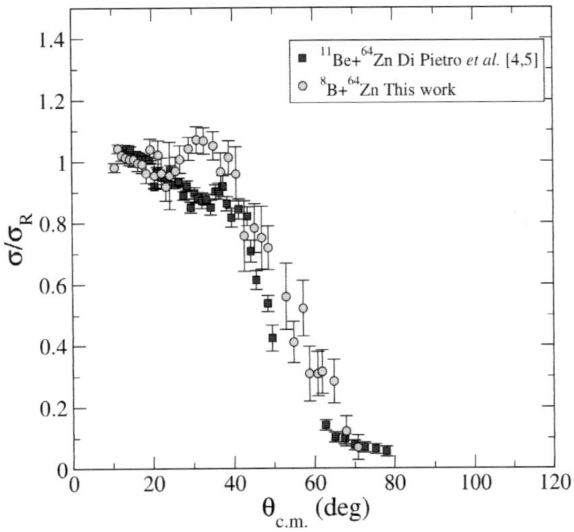

Figure 3. (Color on line) Elastic scattering angular distribution of $^{8}B+^{64}Zn$ (green circles), $^{11}Be+^{64}Zn$ (blue squares) as a ratio to the Rutherford cross-sections. The data of $^{11}Be+^{64}Zn$ are from [4, 5].

The agreement between the experiment and the sum of EBU and NEB cross-section is rather good except in the angular region θ_{lab} <20° where the data are underestimated. From this comparison it can be observed that the dominant reaction mechanism for the ^{7}Be production at forward angles is the EBU. This result is similar to what has been observed in the case of inclusive breakup of the neutron-halo nucleus ^{11}Be and at variance with what observed with ^{6}He on the same target. In this last case the dominant process is the NEB. The total breakup cross-section (EBU+NEB) is much larger in the neutron-halo case than it is in proton-halo.

3. The $^{9}Li+^{4}He$ experiment

The experiment was performed at the TRIUMF (Canada) using a 32 MeV ^{9}Li beam delivered by ISAC-II impinging on a ^{4}He gas target [16]. The Thick Target Inverse Kinematic (TTIK) technique was adopted for the experiment [17]. The ^{9}Li beam was stopped in the gas before reaching the detectors placed at 0° in the laboratory, while the recoiling α particles, owing to the lower stopping power with respect to the beam, were detected. The target consisted of the TUDA chamber filled with isotopically pure ^{4}He gas at pressure of 650 Torr. A Kapton window $\approx$ 12 μm thick was used to separate the gas-filled chamber from the high vacuum beam line. α-particles from elastic scattering were detected and discriminated from other reaction processes using Time of Flight (ToF) techniques. For an extended target, in fact, the ToF measurement allows discrimination of elastic scattering from other reaction events, for example, inelastic scattering. To measure ToF, a microchannel plate (MCP) detector [18] was placed under vacuum, upstream of the Kapton window; it gave a signal whenever a ^{9}Li beam particle entered into the chamber. It was also used for the cross-section normalisation
The detection system consisted of three telescopes each made of 50×50 mm^2, 50 μm thick silicon detector segmented into four-quadrants as ΔE and a 50×50 mm^2 and 1500 μm thick single PAD silicon as residual energy detector.

Monte Carlo simulations were performed to determine the relationship between measured quantities (energies and position of the detected αs) and reconstructed ones (c.m. energy and scattering c.m. angle). The simulations included the energy and angular straggling for the beam

Journal of Physics: Conference Series **2619** (2023) 012007 doi:10.1088/1742-6596/2619/1/012007

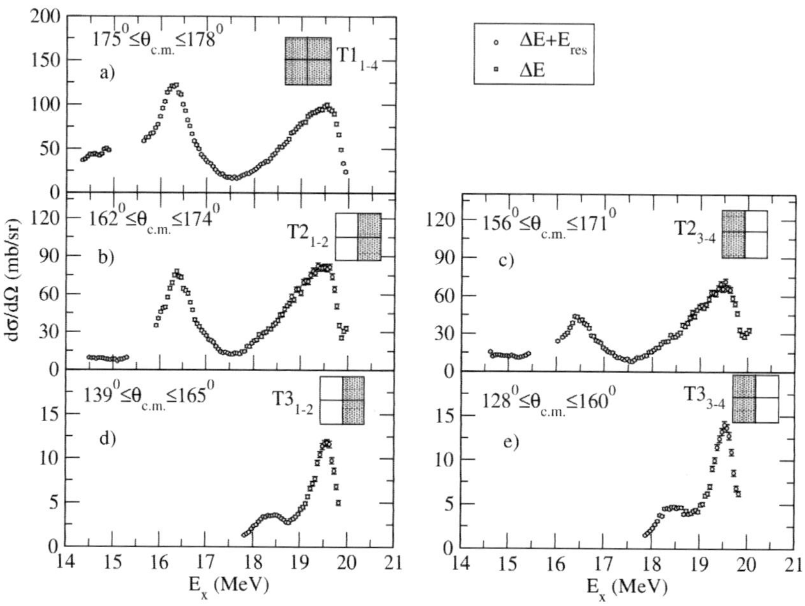

Figure 4. ^{13}B excitation functions. Each plot represents the excitation function in the angular region specified.

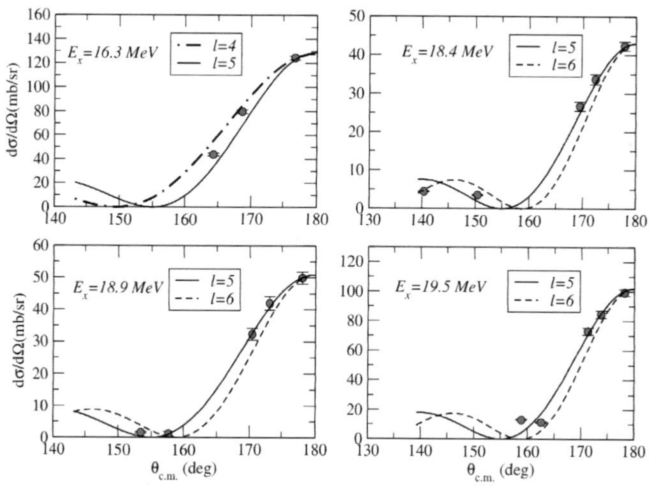

Figure 5. (Symbols) angular distribution at four excitation energies. (dot-dashed line) ℓ=4 (full line) ℓ=5, (dashed line) ℓ=6 Legendre polynomials normalised to the experimental cross-section at the most backward c.m. angle. See text for details.

and the recoiling α-particles.

The ^{13}B excitation function was obtained from the $E_{c.m.}$ spectrum by adding the two-body Q_{val} = 10.82 MeV. In fig. 4 the excitation functions at various detector scattering angles are shown. The cross-section shown in figure 4 presents broad peaks and an angular distribution peaked at backward angles (see fig.5). This behaviour may indicates the occurrence of molecular-type resonances produced by an orbiting-type phenomenon. Above the Coulomb barrier, shape resonances can show-up at angular momentum $\ell \sim kR$. These resonances correspond to quasi-

 doi:10.1088/1742-6596/2619/1/012007

molecular states with angular distributions characteristic of the ℓ [19] and peaked at large angles. This phenomenon is a rolling process where projectile and target touch at the surface at an impact parameter near the sum of their radii; after rolling, the two nuclei break apart. As a result, at backward angles the elastic cross-section can be enhanced. However, for neutron rich nuclei, this surface interaction region will be extended. In the case of the reaction ^{9}Li+^{4}He, the ^{9}Li extended low density neutron field generates a weak attraction for orbital angular momentum components of the wave function corresponding to peripheral interactions between the two ions. In [16] a model, called Transient Rotational Model (TRM), was developed to explain this phenomenon. The various parameters of the model where deduced from Kanada Enyo et al. [10]. It was found that the model predicts values reasonably close to the energies and cross section magnitudes of the 16.4 and 19.5 MeV resonances. This supports the underlying assumption of the model that quasi-molecular states are being formed and those states are originated by the weak absorption at the surface due to the neutron-skin present in ^{9}Li.

4. Summary

This contribution presents a discussion about the behaviour, in low-energy nuclear reactions, of proton-halo nuclei with respect to neutro-halo. The ^{8}B+^{64}Zn elastic scattering angular distribution, unlike the neutron-halo case, shows the presence of the Coulomb-nuclear interference peak. The CDCC calculations showed that the reaction dynamics for the proton-halo ^{8}B exhibits small effects of coupling to the continuum and the total $\sigma_{\rm R}$ is similar to that of stable weakly bound nuclei on the same target.

At small angles in the ^{7}Be inclusive angular and energy distributions dominates the elastic breakup process, whereas at larger angles non-elastic breakup becomes important. This behaviour is similar to what found in the breakup of ^{11}Be. The total breakup cross-section is however smaller in the proton-halo case.

The second part of this contribution is dedicated to the investigation of the formation of possible molecular resonances that are predicted to exist in neutron-rich B-isotopes.

In particular in [10] it is predicted the possibility that the ^{13}B nucleus could exhibit molecular type resonances above the α threshold with ^{4}He and ^{9}Li clusters coupled in rotational states at high excitation energies. The ^{4}He+^{9}Li elastic excitation function measurement is one of the most direct way to study these types of resonances. The results show the presence of broad resonances at energies approximately 16.5 and 19.5 MeV. To interpret these data a simplified molecular rotational reaction model was developed. It was found that the model predicts reasonably well the energies and cross sections of the 16.4 and 19.5 MeV resonances, supporting the idea that quasi-molecular states are being formed and those states are originated by the weak absorption at the surface due to the neutron-skin present in ^{9}Li.

5. Acknowledgments

This work has been supported by the European Commission Horizon 2020 framework trans-national access program of the EU project ENSAR2 (Grant no. 654002).

References

[1] A.M. Sánchez-Benítez et al. Study of the elastic scattering of ^{6}He on ^{208}Pb at energies around the Coulomb barrier. *Nucl. Phys. A*, 803(1):30 – 45, 2008.

[2] A. Di Pietro et al. Reactions induced by the halo nucleus ^{6}He at energies around the Coulomb barrier. *Phys. Rev. C*, 69:044613, Apr 2004.

[3] M. Cubero et al. Do halo nuclei follow Rutherford elastic scattering at energies below the barrier? The case of ^{11}Li. *Phys. Rev. Lett.*, 109:262701, Dec 2012.

[4] A. Di Pietro et al. Elastic scattering and reaction mechanisms of the halo nucleus ^{11}Be around the Coulomb Barrier. *Phys. Rev. Lett.*, 105:022701, Jul 2010.

[5] A. Di Pietro et al. Experimental study of the collision ^{11}Be + ^{64}Zn around the Coulomb barrier. *Phys. Rev. C*, 85:054607, May 2012.

[6] R. Raabe et al. No enhancement of fusion probability by the neutron halo of ^{6}He. *Nature*, 431:823 – 826, 2004.

[7] J.P. Fernández-García et al. ^{11}Li breakup on ^{208}Pb at energies around the Coulomb Barrier. *Phys. Rev. Lett.*, 110:142701, Apr 2013.

[8] V. Guimarães et al. Nuclear and Coulomb interaction in ^{8}B breakup at sub-Coulomb energies. *Phys. Rev. Lett.*, 84:1862–1865, Feb 2000.

[9] M. Mazzocco et al. Elastic scattering for the ^{8}B and ^{7}Be+^{208}Pb systems at near-Coulomb barrier energies. *Phys. Rev. C*, 100:024602, Aug 2019.

[10] Yoshiko Kanada-En'yo, Yusuke Kawanami, Yasutaka Taniguchi, and Masaaki Kimura. Cluster States in 13B. *Progress of Theoretical Physics*, 120(5):917–935, 11 2008.

[11] R. Spart, A. Di Pietro, P. Figuera, O. Tengblad, A.M. Moro, I. Martel, J.P. Fernndez-Garca, Jin Lei, L. Acosta, M.J.G. Borge, G. Bruni, J. Cederkll, T. Davinson, J.D. Ovejas, L.M. Fraile, D. Galaviz, J. Halkjaer Jensen, B. Jonson, M. La Cognata, A. Perea, A.M. Snchez-Bentez, N. Soi, and S. Vials. Probing proton halo effects in the 8b+64zn collision around the coulomb barrier. *Physics Letters B*, 820:136477, 2021.

[12] G. Marqunez-Durn, L. Acosta, R. Berjillos, J.A. Dueas, J.A. Labrador, K. Rusek, A.M. Snchez-Bentez, and I. Martel. GLORIA: A compact detector system for studying heavy ion reactions using radioactive beams. *Nucl. Instr. and Meth. A*, 755:69 – 77, 2014.

[13] E.F. Aguilera et al. Reaction cross sections for ^{8}B, ^{7}Be, and ^{6}Li+^{58}Ni near the coulomb barrier: Proton-halo effects. *Phys. Rev. C*, 79:021601, Feb 2009.

[14] A. García-Camacho, G. Blanchon, A. Bonaccorso, and D. M. Brink. All orders proton breakup from exotic nuclei. *Phys. Rev. C*, 76:014607, Jul 2007.

[15] J.F. Liang et al. Dynamic polarization in the coulomb breakup of loosely bound ^{17}f. *Phys. Lett. B*, 681:22, 2009.

[16] A. Di Pietro, A.C. Shotter, J.P. Fernndez-Garca, P. Figuera, M. Fisichella, A.M. Moro, M. Alcorta, M.J.G. Borge, T. Davinson, F. Ferrera, A.M. Laird, M. Lattuada, N. Soi, O. Tengblad, D. Torresi, and M. Zadro. Hints of quasi-molecular states in 13b via the study of 9li-4he elastic scattering. *Physics Letters B*, 832:137256, 2022.

[17] V. Z. Goldberg, G. V. Rogachev, W. H. Trzaska, J. J. Kolata, A. Andreyev, C. Angulo, M. J. G. Borge, S. Cherubini, G. Chubarian, G. Crowley, P. Van Duppen, M. Gorska, M. Gulino, M. Huyse, P. Jesinger, K.-M. Källman, M. Lattuada, T. Lönnroth, M. Mutterer, R. Raabe, S. Romano, M. V. Rozhkov, B. B. Skorodumov, C. Spitaleri, O. Tengblad, and A. Tumino. Investigation of the α-cluster structure of ^{22}Ne and ^{22}Mg. *Phys. Rev. C*, 69:024602, Feb 2004.

[18] A. Musumarra, P. Figuera, F. De Luca, A. Di Pietro, P. Finocchiaro, M. Fisichella, M. Lattuada, A. Pakou, M.G. Pellegriti, G. Randisi, G. Scalia, C. Scir, S. Scir, V. Scuderi, D. Torresi, and M. Zadro. Measuring total reaction cross-sections at energies near the coulomb barrier by the active target method. *Nucl. Instr. Methods A*, 612(2):399–406, 2010.

[19] K. W. McVoy. Regge Poles and Strong Absorption in Heavy-Ion and α-Nucleus Scattering. *Phys. Rev. C*, 3:1104–1117, Mar 1971.

Analysis of α + ^{16}O elastic scattering in the semi-microscopical dispersive optical model

S. Gomez-Rivera, T. L. Belyaeva

Abstract. We study the elastic scattering cross sections of the $\alpha + ^{16}$O system at 49.5, 69.5, and 80.7 MeV using the optical potentials calculated in the framework of the semi-microscopical dispersive optical model. In this model, the optical potential contains an energy-independent part calculated in the double folding model with a realistic effective nucleon-nucleon interaction, as well as an energy-dependent dynamic polarization potential, the parameters of which are adjusted to the data. The parameters of the real and imaginary parts of the dynamic polarization potential are connected by the dispersion relations.

Universidad Autonoma del Estado de Mexico, Av. Instituto Literario 100, C.P. 50000, Toluca, Mexico
Email: sgomezr002@alumno.uaemex.mx

1. Introduction

The search for an effective potential that describes the interaction of colliding nuclei in the elastic channel is the important part of heavy-ion reaction studies. This problem is usually solved by using an optical model [1-3], where an effective complex nucleus-nucleus potential or optical potential (OP) is constructed phenomenologically and contains various parameters determined by fit the data. Also the global OPs that have few energy-dependent parameters are deduced to describe the interaction of light projectiles with target nuclei at wide energy range [2, 4].

Alternatively, a semi-microscopic approach involves the effective OP, which is represented as a sum of two components. The first of them, a static mean-field potential, is commonly calculated in the microscopic double-folding model [3,5-13] by using a realistic effective nucleon-nucleon interaction [14]. The second part is usually called a dynamic polarization potential (DPP); it depends on the projectile energy and describes the inelastic interaction between colliding nuclei. The real and imaginary parts of the DPP are constructed phenomenologically. Thus a semi-microscopic approach combines a microscopic calculation of the mean-field potential and empirical adjustment of the DPP. The effective nucleus-nucleus potential must be nonlocal, complex, and depends on the nuclear matter density [3]. The semi-microscopic OP can be written in the following form [10, 11]:

$$U_E = V_F + V_P + iW + V_C \tag{1}$$

where V_F is a mean-field potential, V_P is the real part of the DPP, W is the imaginary part of the DPP, and V_C is the Coulomb potential.

2. Calculation of the double folding potential

The mean-field potential for the interaction of two nuclei is expressed as a sum of the direct V_F^D and exchange V_F^E components,

$$V_F = V_F^D + V_F^E, \tag{2}$$

and is written in terms of folding integral by considering nuclei in their ground states [6],

$$V_F = < \phi_{p_0}\, \phi_{t_0}\, |V|\, \phi_{p_0}\, \phi_{t_0}\ >, \tag{3}$$

where V denotes the interaction of the projectile p and the target t, which is presented as a sum of two-body interactions

$$V = \sum_{pt} v_{pt}. \tag{4}$$

Involving the exchange of nucleons, an extra term with the exchange operator of all coordinates of the two interacting nucleons is added

$$v_{pt} \to v_{pt}(1 - \widehat{P}_{pt}).$$

Thus, the direct and exchange terms of Eq. (2) have the following explicit forms [8-12]

$$V_F^D(\mathbf{r}) = \int \rho_p(\mathbf{r}_p)\rho_t(\mathbf{r}_t)v^D(\rho, E, r_{pt})\ d^3r_pd^3r_t, \tag{5}$$

$$V_F^E(\mathbf{r}) = \int \rho_p(\mathbf{r}_p, \mathbf{r}_p + \mathbf{r}_{pt})\rho_t(\mathbf{r}_t, \mathbf{r}_t - \mathbf{r}_{pt})v^E(\rho, E, r_{pt})\exp\left(\frac{i\,\mathbf{K}(r)\mathbf{r}_{pt}}{M}\right)\ d^3r_pd^3r_t \tag{6}$$

in terms of the nuclear matter distribution functions $\rho_{p,t}(r)$ of each of the colliding nuclei, the effective density-dependent nucleon-nucleon interaction potential $v^{D,E}(E, \rho, r_{pt})$, and the local momentum

$$K(r) = \sqrt{\frac{2\mu}{\hbar^2}\,[E - V(r) - V_C(r)]},$$

where $\mathbf{r}_{pt} = \mathbf{r} - \mathbf{r}_p + \mathbf{r}_t$.

The most important part of the calculation of the mean-field potential is a choice of the realistic effective nucleon-nucleon (NN) interaction. We use in our calculations the M3Y-Paris interaction [6, 9, 10, 11, 12] (corresponding to the notation DDM3Y1(Paris)), in which the nucleon knock-on exchange is included and then a density-dependent correction is performed following Refs. [9, 11]:

$$\begin{aligned} v^D(\rho, E, r_{pt}) &= F(\rho, E)v_{NN}^D(r_{pt}), \\ v^E(\rho, E, r_{pt}) &= F(\rho, E)v_{NN}^E(r_{pt}), \end{aligned} \tag{7}$$

where

$$F(\rho, E) = C[1 + Ae^{-B\rho} + G\rho], \tag{8}$$

$$v_{NN}^D(r) = 11061.625\frac{e^{-4r}}{4r} - 2537.5\frac{e^{-2.5r}}{2.5r}, \tag{9}$$

and

$$v^E(r) = -1524.25\frac{e^{-4r}}{4r} - 518.75\frac{e^{-2.5r}}{2.5r} - 7.8474\frac{e^{-0.7072r}}{0.7072r}. \tag{10}$$

Parameters used at the density-dependent function $F(\rho, E)$ are shown in Table 1.

Table 1. Parameters of the density-dependent function $F(\rho, E)$ for the DDM3Y (Paris) potential

Interaction	C	A	B (fm^3)	G (fm^3)
DDM3Y1(Paris)	0.2963	3.7231	3.7384	0.0

The nuclear matter densities of the projectile and target are taken based on the 3-parameter modified Fermi form [15]

$$\rho(r) = \rho_0 \left(1 + \frac{wr^2}{c^2}\right) \left[1 + \exp\left(\frac{r-c}{z}\right)\right].$$

For α particles, we have the following values: $c_p = 1.008$ fm, $z_p = 0.327$ fm, $w_p = 0.445$, likewise, for the target, these parameters are chosen as: $c_t = 2.608$ fm, $z_t = 0.513$ fm, $w_t = -0.051$ [15].

3. The DPP potential

The dynamic polarization potential ($V_P + iW$) describes virtual excitations to inelastic channels and the flux loss (absorption) in the elastic channel. The DPP is a phenomenological potential and is obtained by comparison with the scattering data.

Following Ref. [11], we present the volume and surface parts of the DPP in the Woods-Saxon form and its derivative

$$V_P(r, E) = \alpha(E) W_0(E) f(x_W) + \beta(E) \times 4W_D(E) \frac{f(x_D)}{dr}, \tag{11}$$

$$W = -W_0(E) f(x_W) + 4W_D(E) \frac{df(x_D)}{dr}, \tag{12}$$

where $f(x) = (e^x + 1)^{-1}$ with $x_{W,(D)} = (r - R_{W,(D)})/a_{W,(D)}$ and $R_{W,(D)} = r_{W,(D)} A^{1/3}$.

Phenomenological parameters $\alpha(E)$, $\beta(E)$, $W_0(E)$, $W_D(E), r_{W\,(D)}$, and $a_{W\,(D)}$ are chosen by fit the data.

4. Results and discussion

In modern studies of nucleus-nucleus elastic scattering, various models of double folding potentials and dynamic polarization potentials are widely used (see, for example, [16-19] and references therein). We study the α + ^{16}O elastic scattering at 49.5, 69.5, and 80.7 MeV. The double folding potentials were calculated with the computer code FOLDEG [11] and the DPPs were calculated with an original code WDPP.

Parameters of the DPP for the α + ^{16}O scattering are shown in Table 2.

Table 2. Parameters of the dynamic polarization potential for the α + ^{16}O scattering

$E_{\rm lab}$[MeV]	$\alpha(E)$	$\beta(E)$	$-W_0$[MeV]	$-W_D$[MeV]	$-J_V$ [MeV·fm^3]
49.5	−0.0001	−0.34	11.9	5.5	416.8
69.5	−0.27	−0.36	13.1	6.4	397.7
80.7	−0.45	−0.33	15.2	4.5	391.0

The geometrical parameters of the DPP were fixed at the values: $r_W = 0.9$ fm, $a_W = 0.45$ fm, $r_D = 1.0$ fm, $a_D = 0.55$ fm.

The calculated nucleus-nucleus potentials are shown in Figure 1, where the mean-field potential, the real part of the DPP, V_P, and the absorption potential W are indicated for each energy.

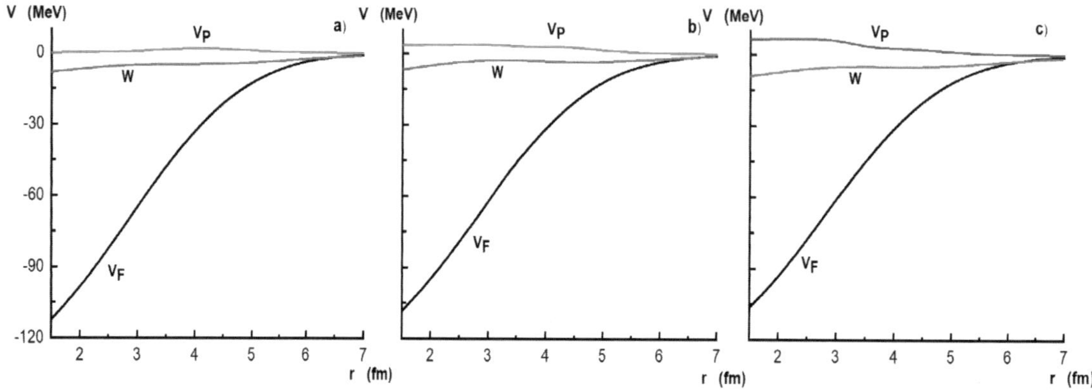

Figure 1. Effective α + ^{16}O potentials resulting from the calculations in the semi-microscopic model at a)$E_{\rm lab}$ = 49.5 MeV, b)$E_{\rm lab}$ = 69.5 MeV, c)$E_{\rm lab}$ = 80.7 MeV

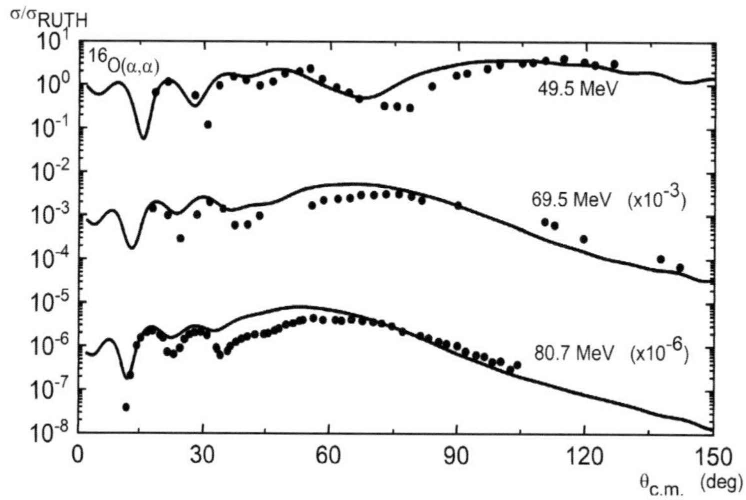

Figure 2. Comparison of the calculated differential cross section of α + ^{16}O elastic scattering with at 49.5, 69.5, and 80.7 MeV

The folding potential V_F is deeper compared to the other parts of the potential, being the dominant part of the effective potential. The real part of the DPP contributes as an energy-dependent correction of the V_F and its influence increases with increase of the projectile energy.

The elastic scattering cross sections calculated using the code FRESCO [20] are compared with the data [4,21] in Fig. 2. The calculated cross sections are in a qualitative agreement with the data and can be improved by applying a tuned version of the model.

5. Conclusion

The semi-microscopic optical model was used for the analysis of the elastic α+ ^{16}O scattering at 49.5, 69.5, and 80.7 MeV. The effective optical potential in this model contains two parts: the first one is a mean-field potential for the interaction of α-particle with ^{16}O calculated within the double-folding model with allowance for exchange effects. The second term is a phenomenological dynamic polarization potential, the real and imaginary parts of the which are related by dispersion relations and are chosen to be proportional adjusted scaling parameters. Their energy independence was controlled by the energy dependence of the DPP volume integrals at fixed geometric parameters.

The mean-field potential was calculated applying the density-dependent version of the M3Y (Paris) effective nucleon–nucleon interactions and a nuclear matter density of the 3-parameter modified Fermi form.

The calculated (using computer code FRESCO) differential cross sections for the elastic $\alpha+$ ^{16}O scattering were compared with the data and a reasonable description of the data was demonstrated. We confirmed that the double folding model with a realistic effective nucleon-nucleon interaction and the exchange term to account antisymmetrization of the system has an advantage to reduce ambiguities inherent to pure phenomenological methods. In general, the applied model involves a smaller number of parameters and makes it possible to control them in accordance with physical conditions.

References

[1] Hodgson P E 1963 *The Optical Model of Elastic Scattering,* Carendon Press, Oxford
[2] Perey C M, Perey F G 1976 Atom. Data Nucl. Data Tables **17** 1
[3] Satchler G R 1983 *Direct Nuclear Reactions,* Clarendon Press, Oxford
[4] Michel F, Albinski J, Belery P, Delbar Th, Gregoire G, Tasiaux B, Reidemeister G 1983 *Phys. Rev. C* **28** 1904
[5] Dover C B, Vary J P 1973 *Phys. Rev. Lett.* **31** 1510
[6] Satchler G R, Love W G 1979 *Phys. Reports* **55** 183
[7] Abele H, Staudt G 1993 *Phys. Rev. C* **47** 742
[8] Khoa D T, von Oertzen W 1993 *Phys. Lett. B* **304** 8
[9] Khoa D T, Satchler G R, von Oertzen W 1997 *Phys. Rev. C* **56** 954
[10] Goncharov S A, Knyaz'kov O M , Kolozhvari A A 1996 *Phys. Atom. Nucl.* **59** 63
[11] Goncharov S A, Izadpanah A 2007 *Phys. Atom. Nucl.* **70** 18
[12] Brandan M E, Satchler G R 1997 *Phys. Reports* **285** 143
[13] Hirabayashi Y, Ohkubo S 2013 *Phys. Rev. C* **88** 014314
[14] Anantaraman N, Toki H, Bertsch G 1983 *Nucl. Phys. A* **398** 269
[15] De Jager C W, De Vries H, De Vries C 1974 *Atom. Data Nucl. Data Tables* **14** 479
[16] Zemlyanaya E V, Lukyanov V K, Kadrev D N, Lukyanov K V, Antonov A N, Gaidarov M K 2020 *J. Phys. Conf. Series 1555 012017*
[17] Durant V, Capel P, Schwenk A 2020 *Phys. Rev. C102, 014622*
[18] Aguilera E F, Torabi F, Morales-Rivera J C 2021 *Phys. Rev. C104, 054612*
[19] Wang K et al. 2021 *Phys. Rev. C104, 054612*
[20] Thompson I J 1988 *Computer Physics Reports* **7** 167
[21] Reed M 1968 Ph. D. thesis, University of California.

Journal of Physics: Conference Series **2619** (2023) 012009 doi:10.1088/1742-6596/2619/1/012009

Developing system arrays for new experimental approach in nuclear astrophysics

G.L. Guardo[1], D. Lattuada[1,2], T. Petruse[3,4] on behalf of ASFIN & ELISSA collaborations

[1] INFN-Laboratori Nazionali del Sud, Catania, Italy
[2] Universitá degli Studi di Enna 'Kore', Enna, Italy
[3] Extreme Light Infrastructure Nuclear Physics/IFIN-HH, Magurele, Romania
[4] Scoala Doctorala de Ingineria si Aplicatiile Laserilor si Acceleratorilor, Universitatea Politehnica, Bucuresti, Romania

E-mail: guardo@lns.infn.it

Abstract. The advent of facilities providing high-intensity and high-resolution gamma ray beams and/or ultra-short and high-repetition laser pulses can potentially open a new path of astrophysical research. Indeed, a pencil size gamma beams with tunable energies from few keV up to tens MeV will offer distinctive chances to conduct precise measurements of small cross sections (on the scale of μb or even smaller) pertaining to nuclear reactions in the field of astrophysics. Consequently, it provides essential data for modeling astrophysical S-factors crucial to stellar evolution. On the other hand, the possibility to mimic the stellar conditions by laser-matter interaction generating a controlled laboratory plasma with thermodynamical status not too different from stellar conditions will open the way for the study of nuclear reactions of utmost importance for nuclear astrophysics.

For photonuclear reactions with astrophysical significance, as photodissociations occur at photon energies slightly above particle emission thresholds due to typical stellar temperatures, the resulting fragments possess low energies spanning from a few hundred keV to a few MeV. Consequently, detectors with low thresholds become imperative in such cases. Also, in the case of laser-induced reactions, in order to detect the fusion products and to measure the laser-accelerated ion distribution a proper system of detection is needed. Depending on the available exit channels of the nuclear reaction of interest, both charged particles and neutrons are foreseen.

Here, we present the Asfin's efforts on developing new detectors arrays suitable for the experimental requirements in these challenging measurements. Indeed, an experimental campaign is ongoing in order to test the feasibility of excitation functions and angular distributions determinations using versatile silicon strip arrays (namely LHASA and/or ELISSA). Moreover, extensive studies and simulations will be presented regarding the developing of a dedicated detection system comprising a cryogenically cooled supersonic nozzle, an appropriate interaction chamber, an array of neutron and charged particle detectors and two compact ion spectrometers for performing systematic study of laser-induced nuclear fusion reactions.

1. Introduction

Photodissociation reactions are of great importance in nuclear astrophysics. They involve the breaking of a nucleus into two or more fragments by the absorption of a photon. These reactions are crucial in the synthesis of heavy elements in the universe, as they can lead to the production of lighter elements that can then undergo additional fusion reactions to form heavier elements [1].

Among the others, in stellar explosive scenarios like type II supernovae, the p-process stands as a significant mode of nucleosynthesis responsible for generating stable nuclides beyond iron that are highly depleted in neutrons. These neutron-deficient nuclides are unattainable through the *s*- and *r*-processes. These particular nuclei are denoted as *p*-nuclei, with the designation "p" indicating their higher proton count compared to other stable isotopes of the same element [1]. As a group, the *p*-nuclei are the rarest among the stable nuclides. It is thought that these nuclides are synthesized by photodisintegration of pre-existing *s*- and *r*-nuclei. There are 35 nuclides classified as the *p*-process nuclides ranging from ^{74}Se to ^{196}Hg that are bypassed by the *s*- and *r*-process [2, 3]. Except for a few cases, there is a lack of data regarding photodissociation reactions that result in the creation of *p*-nuclei, necessitating the use of Hauser-Feshbach calculations. However, the final abundances of *p*-nuclei obtained from type II supernovae are highly sensitive to different prescriptions of the Hauser-Feshbach model. This sensitivity introduces significant systematic errors, potentially causing inaccurate and misleading comparisons between models and observational data. As a result, pinpointing the correct site for the *p*-process and comprehending *p*-process nucleosynthesis becomes challenging [4].

For example, *p*-nuclei of Mo and Ru are produced only by *p*-process in the scenario of supernova explosion. Astrophysical simulations indicate a consistent underproduction in the nucleosynthesis of $^{92;94}$Mo and $^{96;98}$Ru when comparing predicted abundances to their observed counterparts in the Solar System composition. Addressing this unresolved issue in *p*-process nucleosynthesis is anticipated through experimental determination of astrophysical reaction rates around Mo and Ru. [5].

Other important photodissociation reactions include the photodissociation of light nuclei such as deuterium, helium-3 and lithium-7. These reactions can contribute to the production of heavier elements through subsequent fusion reactions [6]. Additionally, photodissociation reactions can be used to study the properties of atomic nuclei, such as their energy levels and decay modes, providing valuable information for nuclear physics research [7].

Overall, photodissociation reactions are a crucial aspect of nuclear astrophysics, playing a key role in the production of the elements and providing a valuable tool for studying the properties of atomic nuclei. With the development of new technological facilities capable of generating high-intensity and high-resolution gamma ray beams, as well as ultra-short and high-repetition laser pulses, there is a potential for a new era of astrophysical research to be opened up [8, 9].

From another point of view, low-energy nuclear physics plays a critical role in two major fields of study: astrophysics and applied (plasma) physics. In astrophysics, it is crucial in determining the primordial abundances in Big Bang Nucleosynthesis (BBN) models, while it can also help in explaining the abundance of elements in the universe [10]. In applied physics, low-energy nuclear physics is significant because it operates in the energy range that is of interest for the design and operation of future fusion power plants [11]. Over the years, both direct and indirect measurements of the cross-sections for these reactions have been conducted. Some of these measurements propose that the presence of electrons, leading to a screening potential, can reduce the Coulomb barrier between the projectile and target nuclei at extremely low energies. This reduction results in an enhanced cross-section compared to interactions with bare nuclei and those occurring in astrophysical plasmas [10, 12, 13, 14, 15, 16, 17, 18, 19, 20, 21, 22, 23, 24].

A different approach is to mimic the stellar conditions by generating a controlled laboratory plasma with thermodynamical status not too different from stellar conditions. This paradigm has long been obstructed by the lack of proper technology, but with the advent of new techniques in laser amplification, many laboratories are starting to study how to create plasmas with temperatures and pressures close to those that exist only in the cores of stars [25, 26].

Considering these fields of experimental opportunities, challenges often arise when conducting

Journal of Physics: Conference Series **2619** (2023) 012009 doi:10.1088/1742-6596/2619/1/012009

experiments under demanding conditions. In the next paragraphs, we aim to showcase the efforts of Asfin in developing new detector arrays specifically designed to meet the requirements of these challenging measurement conditions and highlight their potential to enhance the accuracy and efficiency of experimental measurements in these challenging environments.

2. Nuclear Astrophysics with γ beam

In a general context, the study of photodissociation reactions is relatively more straightforward due to the phase space factor amplifying the (γ,p) and (γ,α) cross sections compared to the inverse process. This is particularly true when a high-quality gamma beam, such as the one available at the HIγS facility or soon to be accessible at the ELI-NP facility, is employed. For photonuclear reactions of astrophysical relevance, photodissociations occur at photon energies slightly above particle emission thresholds due to the typical temperatures in stars. Consequently, the resulting fragments have low energies ranging from a few hundred keV to a few MeV. Moreover, to gain a comprehensive understanding of the reaction mechanism, it is crucial to measure excitation functions and angular distributions over a wide range. Considering the potential presence of beam-induced background at forward angles, it is imperative to achieve the broadest possible angular coverage to ensure a sufficiently large range for effective nuclear spectroscopy.

As most reactions involve charged particles in the exit channel, the use of low-threshold detectors is essential. However, conventional particle identification techniques, like measuring energy loss in a thin detector or employing pulse shape analysis, have proven currently unattainable. Additionally, the adoption of a compact device, while advantageous for easy integration with ancillary detectors such as neutron arrays, renders the use of time-of-flight (ToF) for particle identification unfeasible. Given these limitations, kinematical identification emerges as the sole viable option presently, as highlighted by [27]. This approach necessitates high angular and energy resolution to effectively discriminate between various reaction kinematics, enabling the separation of the reaction of interest from others.

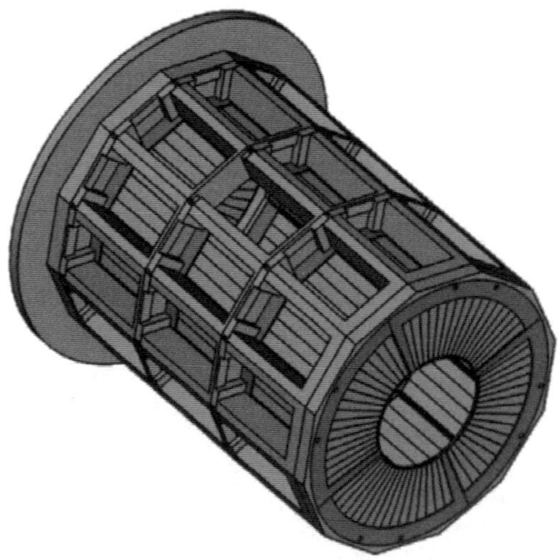

Figure 1. Full design of the ELISSA detector. It will be made of n. 3 rows of PSD detectors to form a barrel plus n. 8 end cap detectors to extend the angular coverage.

For all these reasons, silicon detectors represent one of the best solution for the construction of a dedicated array for a number of reasons: (i) they guarantee exceptional energy resolution, (ii) they have almost 100% efficiency in charged particle detection, (iii) thresholds can be set to very small values, (iv) they are not sensitive or very little sensitive to neutrons, gamma rays, electrons and positrons, which would constitute the beam induced background. Finally, silicon strip detectors would couple enhanced angular and energy resolution with high detection efficiency [28, 29]. High granularity is not a crucial aspect as a low counting rate is expected, making the probability of multiple hit very small.

Thus, the Extreme Light Infrastructure Silicon Strip Array (ELISSA) is under development in a common effort between the Asfin group from LNS-INFN and the Nuclear Astrophysics group of ELI-NP [30]. The full array will consist of 3 rings of 12 X3 position-sensitive detectors produced by Micron Semiconductor Ltd. [31], in a barrel like configuration, ensuring a total angular coverage in the laboratory system of about 100°. The angular coverage is extended down to about 20° (160° at backward angles) by using end cap detectors such as the assembly of four QQQ3 segmented detectors by Micron Semiconductor Ltd. [31]. A sketch of the final expected setup is shown in Fig. 1. The central ring contains only 11 detector in order to permit the positioning of the target in the middle of the detector holder, so that the total number of X3 detectors is 35. In order to ensure the maximum versatility in terms of physical cases, ELISSA project will provide two different detector thickness: 300μm and 1000μm [32]. The X3 is a 4-strip detector, 4 cm wide, position sensitive along the longitudinal axis (7.5 cm long), leading to a position resolution better than 1mm around 10 MeV, while QQQ3 is a segmented detector with 16 radial segments (3 mm pitch) and 16 angular segments, with inner and outer radii of 4.91 and 10 cm and 1000μm thick.

Another array was already developed at LNS-INFN in Catania (Italy) and it is called Large High-resolution Array of Silicon for Astrophysic (LHASA). LHASA (Fig. 2) consists of six YY1 silicon strip detectors with a thickness of 300 μm mounted in a lamp-shade configuration and optimized to detect alpha particles in a wide angular range.

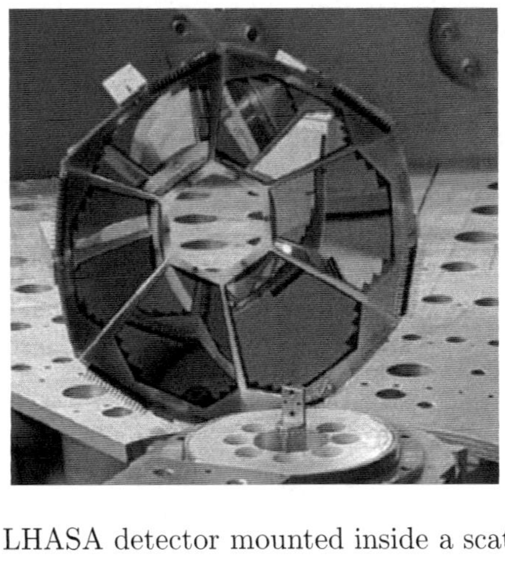

Figure 2. The LHASA detector mounted inside a scattering chamber.

2.1. Experimental test

In the framework of developing, testing and commissioning the silicon arrays described in the previous section, we proposed and studied some direct cross section measurements of interest for nuclear astrophysics and very suitable for pointing out the experimental feature of our detectors. Here in the following a brief description of those experiments:

2.1.1. The ^{7}Li(γ,t)^{4}He cross section measured with a LHASA like detector The ^{7}Li(γ,t)^{4}He ground state cross section was measured for the first time using monoenergetic γ rays at the High Intensity Gamma-ray Source (HIγS) due to its importance for the primordial Li problem [33, 34]. Although over the last 30 years most measurements of the inverse ^{3}H(α,γ)^{7}Li reaction have concentrated in an energy range below E_γ = 3.65 MeV [35, 36, 37, 38], measurements at higher energies could potentially restrict the extrapolation to astrophysically important energies. The experiment was carried out using the large area annular silicon detector array (SIDAR), very similar to the proposed LHASA. A detailed description of the experimental approach, data analysis and astrophysical results can be found here [39]. The results of this experiment clearly prove that photodisintegration cross sections can be determined with good accuracy using a large-area silicon detector array, pushing for the development of ELISSA and LHASA detectors.

2.1.2. Direct measurement of the ^{19}F(p,α)^{16}O reaction cross section The ^{19}F(p,α)^{16}O reaction is an important destruction channel in Asymptotic Giant Branch (AGB) stars and plays a significant role in the nucleosynthesis of heavy elements in the Universe [40]. To better understand the astrophysical implications of this reaction, an experimental campaing was conducted using ELISSA coupled with LHASA, forming a silicon detector array with high angular coverage and extraordinary angular and energy resolution. A first experiment was performed at INFN – Laboratori Nazionali del Sud, Catania (Italy) using the 15 MV Van der Graaf Tandem that provided a ^{19}F beam with an energy range from 9 up to 18.5 MeV. A comprehensive account of the experimental method, analysis of the data, and astrophysical findings can be retrived here [41]. A second experiment focused on the ^{19}F(p,α_π)^{16}O and ^{19}F(p,α_γ)^{16}O reaction channels was recently performed at IFIN-HH, Magurele (Romania) using the 3 MV Tandetron accelerator that provided a ^{19}F beam spanning an energy range from 7 to 15 MeV with a similar approach of the previous experiment. In this case the detection setup consisted of one ring of ELISSA (12 Micron X3 position sensitive silicon strip detectors of 1000 μm thick) and LHASA (5 Micron YY1 silicon strip detectors of 300 μm thick [31]). The high precision of the ELISSA and LHASA detector system will allow for a detailed study of the angular distribution of the emitted α particles, which provides insights into the reaction mechanism and the factors that determine the cross section. The analysis is still ongoing.

3. Nuclear Astrophysics with laser beam

The study of nuclear reactions in laboratory has always been hindered by the very low cross-sections values at energies of astrophysical interest (1-100keV). This leads nuclear astrophysicists either to build huge and expensive underground laboratories where to perform long experiments with low and controlled background (LUNA [42], JUNA [43]), or to exploit indirect methods usually involving nuclear-structure models such as the Trojan Horse Method (THM [44]) or Asymptotic Normalization Coefficient (ANC [45]). Both approaches have proven to be successful to a great extent; nevertheless, plasma in stellar objects is a very different state from the solid or gas targets commonly used in standard nuclear physics experiments involving conventional accelerators. It is well known that the plasma state affects in a non-negligeable way many nuclear processes, such as electron screening in fusion reactions and Non-Local Thermodynamic Equilibrium effects in beta decays.

Our current knowledge on laser-matter interaction, nurtured by the recent developments in laser pulse amplification, provides a scenario where the interplay between the laser intensity (in the regime above 10^6 W/cm^2) and the target composition (thickness and structure) holds a key role in determining the paradigm of the ion acceleration and its scale [46]. Among the others, the interaction of a material prepared at a thermodynamic state close a critical point (i.e., where matter coexists in different states, thermodynamically close to phase shifts) with lasers shows a more intense interaction with respect to conventional materials such as gas or solid and results in super-hot matter and higher laser absorption, finally forming plasmas with multi-keV temperature alongside strong EMP and x-ray emission [47]. These energies overlap with the typical temperatures of stellar cores where thermonuclear reactions occur, thus making this paradigm a perfect scenario for nuclear astrophysics research.

In particular, the interaction of ultra-short laser pulses with an expanding gas mixture at controlled temperature and pressure inside a vacuum chamber has been demonstrated the formation of molecular clusters that maximizes the laser absorption, causing the full strip of the electrons of the gas [48, 49, 50]. The sudden onset of a positive space charge triggers the Coulomb dissociation of the clusters resulting in multi-keV ion acceleration with a nearly isotropic emission (Coulomb Explosion, CE). The accelerated ions interact in the surrounding plasma and nuclear reactions take place. By measuring the ion energy distribution and detecting the products of the occurred nuclear reactions, it is possible to infer important information on the cross section of the nuclear reactions of interest [51, 52].

3.1. The VALAR project
The systematic study of laser-induced nuclear fusion reactions using the Coulomb Explosion paradigma obviously requires a dedicated system as the one that is now under development at INFN - Laboratori Nazionali del Sud in Catania (Italy). The Versatile Array for Laser-induced Astrophysics Research (VALAR) will consist of (i) a cryogenically cooled supersonic nozzle, (ii) an appropriate interaction chamber, (iii) an array of neutron and charged particle detectors and (iv) two compact ion spectrometers.

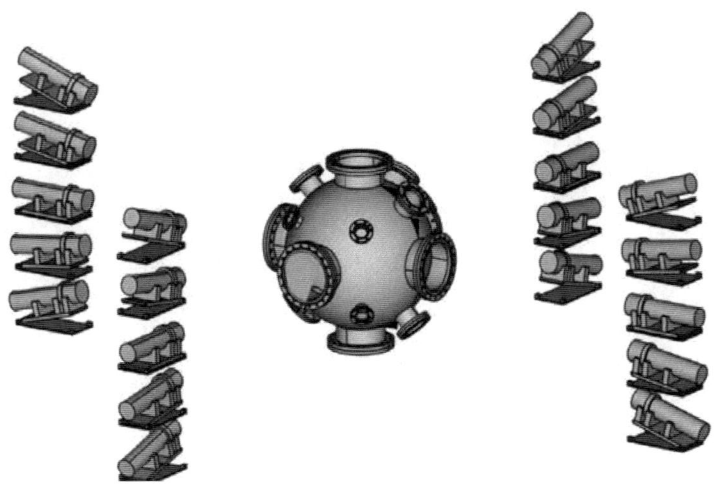

Figure 3. Schematics view of the valar neutron array with prototype supports and chamber.

In particular, the VALAR detection system will be optimized to detect the fusion products and to measure the laser-accelerated ion distribution. Depending on the available exit channels of the nuclear reaction of interest, both charged particles and neutrons are foreseen. For the latter, the use of both plastic and liquid scintillators in Time of Flight (ToF) configuration proved to be an excellent choice owing to their linear and fast response coupled to simple electronics. On the other hand, the detection of both the fusion products and the primary accelerated protons, alpha particles and light ions are obstructed by the high electromagnetic background induced by the interaction of high-power laser pulses with the target together with the concurrent generation of a strong EMP signal. For this reason, fast detectors with high radiation hardness, such as Silicon Carbides (SiC [53]) and Polycrystalline CVD Diamonds [54] in ToF configuration are perfect candidates for charged particle detection. A schematic view of the final configuration of the VALAR array is reported in fig. 3.

3.2. The deuterium-deuterium nuclear fusions as test-case

The deuterium burning has been chosen for the VALAR commissioning thanks to the recent history of successful experiments with deuterium clusters irradiated by high-power laser pulses [55]. Moreover, the importance of deuterium burning is twofold: it is crucial for the Big Bang Nucleosynthesis, since it represents the most crucial reaction in this scenario and it is a reference for the power plants applications. In addition, very few data are available below 15 keV and the inferred electron screening potential from direct measurements is significantly larger than the adiabatic limit [56].

A preliminary experiment was performed using the Texas Petawatt laser (TPW) which delivered 150-270 fs pulses at 1057 nm wavelength and energy ranging from 90 to 180 J to D_2 or CD_4 molecular clusters. The detection setup consisted of five EJ-232Q and EJ-200 plastic scintillation detectors forming a prototype of the final VALAR detection array. In this experimental setup, a highly concentrated laser beam is directed at a gas mixture containing D_2 clusters and ^{3}He atoms. The intense electromagnetic field of the laser temporarily strips away the electrons from the clusters, which are generated during the rapid expansion from the nozzle. This action results in the acceleration of deuterium ions, triggered by the abrupt emergence of the repulsive Coulomb potential attributable to their positive charges. Consequently, ions with multi-keV kinetic energies are generated through this process known as Coulomb explosion. These deuterium ions can collide with each other and generate d+d fusion reactions or they can collide with deuterium atoms at rest in the gas jet outside the focal volume. ^{3}He atoms do not absorb the laser energy efficiently because they do not form clusters at 86 K, but an energetic deuterium ion can collide with a cold ^{3}He atom resulting in d+^{3}He fusion reaction. A detailed description of the data analysis can be found here [57]. By measuring the plasma distribution, its volume, ion concentration, density and the number of fusions occurring for each reaction, in this case we were able to derive the S-factor without any model assumption and compare it with other experiments finding a good agreement with conventional beam-target data within the experimental error. These studies successfully demonstrate the reliability of the technique and opened the way for the study of other reactions of utmost importance for nuclear astrophysics and applied research such as the proton-boron, deuterium-tritium, deuterium-^{3}He, ^{12}C-^{12}C, ^{16}O-^{16}O fusion reactions.

4. Conclusions

The emergence of high-intensity and high-resolution gamma-ray beam facilities presents a significant opportunity for nuclear physics and astrophysics. These facilities enable the measurement of reactions of primary astrophysical significance that were previously inaccessible, marking a pioneering step in advancing our understanding of these phenomena. At the same time, the advent of high-power and high-repetition rate lasers open the way for systematic studies of nuclear astrophysics reaction in plasma with stellar-like conditions.

To catch these new experimental opportunities, the Asfin group is developing new arrays with peculiar characteristics: high-resolution, low-threshold, high-efficiency, compactness, etc. Preliminary results on methods and arrays make us confident for the new perspectives and open the way for new experimental campaigns.

References

[1] C. Iliadis, *Nuclear Physics of Stars*, (Weinheim, Wiley-VCH Verlag, 2007)
[2] T. Rauscher, N. Dauphas, I. Dillmann, C. Fröhlich, Z. Fülöp, and G. Gyürky, 2013 *Rep. Prog. Phys.* **76**, 066201.
[3] M. Pignatari, K. Göbel, R. Reifarth, and C. Travaglio, 2016 *Int. J. Mod. Phys. E* **25**, 1630003.
[4] W. Rapp et al., 2005 *Nucl. Phys. A* **758**, 545.
[5] H.Y. Lan et al., 2022 *Phys. Rev. C* **105** 044618.
[6] M. La Cognata et al., 2017 *J. of Inst.* **12** C03079.
[7] N. Pietralla, J. Isaak, and V. Werner, 2019 *Eur. Phys. J. A* **55**, 237.
[8] D. L. Balabanski et al., 2017 *EPL* **117**, 28001.
[9] D. Filipescu et al., 2015 *Eur. Phys. J.* **A 51**, 185.
[10] A. Tumino et al. 2014, *Astroph. J.* **758**, 96.
[11] H. Hora et al., 2017, *Laser and Particle Beams* **35**, 730.
[12] The First Research Group, 1985 *Chinese J. of Nucl. Phys.* textbf9, 723.
[13] D. S. Leonard, H. J. Karwowski, C. R. Brune, B. M. Fisher, and E. J. Ludwig 2006, *Phys. Rev. C* **73**, 045801.
[14] R. L. Schulte, M. Cosack, A. W. Obst, and J. L. Weil, 1972 *Nucl. Phys. A* **192**, 609.
[15] R. E. Brown and N. Jarmie, 1990 *Phys. Rev. C* **41**, 1391.
[16] U. Greife, F. Gorris, M. Junker, C. Rolfs, and D. Zahnow, 1995 *Z. Phys. A* **351**, 107.
[17] A. Krauss, H. W. Becker, H. P. Trautvetter, C. Rolfs, K. Brand, 1987 *Nucl. Phys. A* **465**, 150.
[18] M. La Cognata et al., 2005 *Phys. Rev. C* **72**, 065802.
[19] M. Aliotta et al., 2000 *Eur.Phys.J. A* **9**, 435.
[20] U. Schroeder, S. Engstler, A. Krauss, K. Neldner, C. Rolfs, E. Somorjai, K. Langanke, 1989 *Nucl. Instr. Meth. B* **40**, 466.
[21] S. Kimura, A. Bonasera, 2005 *Nucl. Phys. A* **759**, 229.
[22] T. D. Shoppa, S. E. Koonin, K. Langanke, R. Seki, 1993 *Phys. Rev. C* **48**, 837.
[23] S. Kimura and A. Bonasera, 2004 *Phys. Rev. Lett.* **93**, 262502.
[24] G. Shaviv and N. Shaviv, 2000 *Astrophys. J.* **529**, 1054.
[25] D. Strickland and G. Mourou, 1985 *Optics Communications*, **55**, 447.
[26] A. J. Kemp, S. C. Wilks, E. P. Hartouni, and G. Grim, 2019 *Nature Communications*, **10**, 4156.
[27] D. Lattuada et al., 2017 *EPJ Web Conf.* **165**, 01034.
[28] S. D. Pain et al., 2007 *Nucl. Instr. Meth. Phys. Res.* **B 261**, 1122.
[29] M. Matos et al., 2010 **PoS NIC XI, 226**.
[30] G. L. Guardo et al., 2017 *EPJ Web of Conf.* **165**, 01026.
[31] Micron Semiconductor, General catalogue, http://www.micronsemiconductor.co.uk/pdf/cat.pdf
[32] S. Chesnevskaya et al., 2018 *J. of Inst.* **13**, T05006.
[33] B. D. Fields, 2011 *Annu. Rev. Nucl. Part. Sci.* **61**, 47.
[34] A. Coc et al., 2014 *J. Cosmol. Astropart. Phys.* **10** 50.
[35] R. H. Cyburt et al., 2008 *JCAP*, **11** 12.
[36] P. Descouvemont et al., 2004 *At. Data Nucl. Data Tables* **88** 203.
[37] C. R. Brune, R.W. Kavanagh and C. Rolfs, 1994 *Phys. Rev. C.* **50** 2005.
[38] T. Neff, 2011 *Phys. Rev. Lett.* **106** 042502.
[39] M. Munch et al., 2020 *Phys. Rev. C* **101**, 055801.
[40] S. Cristallo et al., 2009 *Astroph. J.* **696**, 797.
[41] G. L. Guardo et al., 2023 *Eur. Phys. J. A* **59** 65.
[42] U. Greife et al., 1994 *NIM A* **350** 327.
[43] L. WeiPing et al., 2019 *Proc. IAU* **15** 313.
[44] A. Tumino et al., 2023 *Ann. Rev. of Nuc. and Part. Sc.* **71** 345.
[45] R. E. Tribble, C. A. Bertulani, M. la Cognata, A. M. Mukhamedzhanov, and C. Spitaleri, 2014 *Rep. on Prog. Phys.* **77** 106901.
[46] G. A. Mourou, T. Tajima, and S. v. Bulanov, 2006 *Rev. of Mod. Phys.* **78** 309.
[47] T. Tajima, Y. Kishimoto, and M. C. Downer, 1999 *Phys. of Plasmas* **6** 3759.
[48] K. W. Madison, P. K. Patel, M. Allen, D. Price, R. Fitzpatrick, and T. Ditmire, 2004 *Phys. Rev. A* **70** 053201.
[49] W. Bang et al., 2013 *Phys. Rev. Let.* **111** 055002.
[50] M. Barbui et al., 2013 *Phys. Rev. Let.* **111** 082502.
[51] W. Bang et al., 2013 *Phys. Rev. E* **88** 033108.
[52] H. J. Quevedo et al., 2018 *Phys. Let. A* **382** 94.
[53] S. Tudisco et al., 2018 *Sensors* **18** 2289.
[54] M. Cipriani et al., 2019 *J. of Inst.* **14** C01027.

Journal of Physics: Conference Series **2619** (2023) 012009 doi:10.1088/1742-6596/2619/1/012009

[55] T. Ditmire, J. Zweiback, V. P. Yanovsky, T. E. Cowan, G. Hays, and K. B. Wharton, 1999 *Nature* **398** 489.
[56] Y. Xu, K. Takahashi, S. Goriely, M. Arnould, M. Ohta, and H. Utsunomiya, 2013 *Nuc. Phys. A* **918** 61.
[57] D. Lattuada et al., 2016 *Phys. Rev. C* **93** 045808.

Monoisotopic targets production from atomic and molecular positive ion implantation in a tandem accelerator at low energies

J. Mas-Ruiz, D. J. Marín-Lámbarri, S. Sandoval-Hipólito, E. Monroy-Díaz, G. Reza, A. O. Valdéz-Guerrero, L. Acosta, E. Andrade, D. Belmont, A. Huerta, C. Solís and E. Chávez

Instituto de Física, Universidad Nacional Autónoma de México, Av. Universidad 3000. 04510, Cd. de México, México.

E-mail: `javier.masruiz91@gmail.com`

Abstract. We produced monoisotopic targets of ^{13}C and ^{28,29}Si with 10^{15-16} atoms/cm^2 areal density by ion implantation in different matrices at the Accelerator Mass Spectrometry (AMS) facility in Mexico. The tandem accelerator terminal voltage was set at the lowest achievable value (200 kV) that yield beams with energy of 435 keV. It is possible to do ion implantation at lower energies in our facility through the use of molecular beams, so in this work we looked for, and successfully found, bound positively charged molecular beams, although with very low intensity. We inject 235 keV C_6^- molecules into the charge of exchange canal (argon) of our tandem accelerator and detected the survival fraction of $C_{2,3,4}$ molecules in 1^+ and 2^+ charge states or above. Our results showed that some molecules remain bound, with one or more electrons missing, after being forced through a low density gas stripper at relatively high energy. In the production process of negative molecular fluxes in the Source of Negative Ions by Cesium Sputtering (SNICS) we found an odd-even effect in their intensity that we understand as the coupling between the fragment size distribution of a dynamic fragmentation of graphite by cesium impact and the electron affinity of the molecules with n-carbon atoms. The availability of these kind of high energy molecular beams opens the door for discovery both in the study of the physics of negative and positive molecular ions.

Keywords: Molecular beams, implantation, tandem accelerator

1. Introduction

In Mexico, LEMA (Laboratorio de Espectrometría de Masas con Aceleradores) is the only AMS (Accelerator Mass Spectrometry) facility and has been described previously in [1, 2, 3]. For general references on the AMS technique see [4, 5]. Since its inauguration in 2013 a prolific production has been achieved through the precise measurement of the ultra low concentration of naturally occurring radioisotopes in matter (^{10}Be, ^{14}C and ^{26}Al). Thousands of samples have been analyzed, some of the more relevant results in applied and interdisciplinary research produced in this facility can be found in

Journal of Physics: Conference Series **2619** (2023) 012010 doi:10.1088/1742-6596/2619/1/012010

[6, 7, 8, 9, 10, 11, 12]. This facility has also been useful for fundamental nuclear physics research [13, 14, 15, 16, 17].

As in every AMS facility, the subject of molecular beams is transcendental. The whole strategy of extracting negative ions from the material in study followed by a reacceleration in a tandem type accelerator, has the main goal to keep under control the molecular flow. In routine AMS work, the survival of molecules of mass 14 u, in positive charge states is of the greatest importance. Previous studies have made possible the development of lower energy systems [18].

Our facility is equipped with a Source of Negative Ions by Cesium Sputtering (SNICS) [19, 20, 21] and a 1 MV tandem type accelerator from High Voltage Engineering Europa (HVEE) [22] which is a modern version of the original idea of Cockroft and Walton [23, 24] in tandem configuration with gas stripper (argon).

We extracted C_n^- (n=1-6) molecules from the SNICS and injected the C_6^- beam into the accelerator, keeping the terminal at the lowest possible voltage (200 kV). Similar molecular beams have been studied before [25, 26, 27], using SIMS (Secondary Ion Mass Spectrometry) with cesium beams of a wide range of energies.

Molecular beams are routinely used to study their intrinsic properties and their interactions with other particles or materials [28, 29]. Most of the time, used beam energies are a fraction of eV. A beam of 10 eV is considered a "high energy" for molecular physics studies. In this work we developed molecular beams that were first formed as negatively charged ions, moving with an initial kinetic energy of 35 keV. Then turned into bound positively charged molecules moving at energies of hundreds of keV. We report on the study of the survival of single and double ionized bound molecules in the high energy side of the tandem accelerator. Previous similar efforts can be found in [30, 31]. These molecular beams can be an option to produce implanted targets for elements that do not form negative ions (like nitrogen, magnesium, etc. Noble gases are not included) [32] and also to implant at lower energies than those attainable for atomic ions.

In the study of nuclear structure (sizes, shapes, quantum numbers) or nuclear dynamics (entrance channel effects, branching ratios, level densities, cluster formation) the use of monoisotopic targets goes from convenient to necessary. Implanted targets are some times a very good option [33, 34, 35, 36, 37, 38, 39, 40, 41, 42, 43].

As part of the low energy nuclear physics program at IFUNAM, we started an effort to produce such monoisotopic targets by ion implantation. In this work we present our first results, where we achieved target thickness of 10^{15-16} ions/cm^2 for ^{13}C and ^{28,29}Si on carbon and aluminum targets as described below.

2. Experimental Methods

The used beams were produced by the 1 MV tandem accelerator facility of the National Laboratory of Accelerator Mass Spectrometry (LEMA) located at the Physics Institute of the National Autonomous University of Mexico. The Nuclear Physics Line (NPL), commissioned at the end of 2017, is an extension to the isotope separator of LEMA that converted the facility into an hybrid one (AMS and ion beam applications, see [3, 44] for a full description and first applications).

2.1. Monoisotopic implanted targets

As part of the low energy nuclear physics program of our group, we are interested in the measurement of ultra low nuclear reaction cross sections. The corresponding experimental setup requires targets with as few contaminants as possible. That calls for high purity elemental materials, and preferably highly enriched in the isotope of interest. For that reason we undertook the development of monoisotopic thin targets by means of ion implantation. A useful reference for "thin" could be those produced by supersonic gas jet flow [45] that achieve areal densities from 10^{16} to 10^{19} atom/cm^2. That is as thin as ten typical atomic layers in solids and as thick as ten thousand. Our facility provides flux of selected isotopes and the idea here is to have them stored in a thin layer within a matrix of a convenient material. In this way, we do produce the monoisotopic target of interest in a fine layer inside a matrix that prevents its quick destruction. We chose thick (millimeter) aluminum and carbon slabs as matrix to be implanted, as they are stubborn enough to tolerate the highest currents we can provide without melting. The energy of the ions to be implanted has to be as low as possible to produce the implanted layer the closest to the beam entrance surface of the matrix. Our tandem accelerator can be operated at voltages as low as 200 keV. So the minimum beam energy is 435 keV, including the 35 keV from the SNICS. With this kinetic energy, the ions are deposited near the surface and deep enough to prevent sputtering when the actual nuclear cross section measurement with high intensity low energy beams are run.

In this article we show the results for the implantation of three different beam matrix combinations. The first one, the easiest one, is the implantation of ^{28}Si on glassy carbon. The ^{28}Si are the most intense beams we can deliver to the scattering chamber [3] and was chosen as the one that could serve as proof of principle test. Other two cases of interest were ran: ^{29}Si on glassy carbon and ^{13}C in aluminum. In all cases the beam energy was 435 keV. Rutherford Backscattering (RBS) analysis were performed on the matrices before and after implantation and compared with SIMNRA [46] simulations to characterize the implanted film.

Figure 1 shows the RBS analysis spectrum obtained with 1.035 MeV protons (data and SIMNRA simulation) of the ^{28}Si implantation test case. In this case the implantation run lasted for 102 minutes. The ^{28}Si film produced inside the glassy carbon matrix is $(1.8\pm5)\times10^{16}$ atoms/cm^2.

^{29}Si was implanted in glassy carbon. Here the beam current's intensity was two orders of magnitude smaller than the previous case, since we used natural silicon cathodes in the ion source. The irradiation time turned out to be too short and the amount of ^{29}Si implanted was too small to be characterized by RBS as in the other two cases.

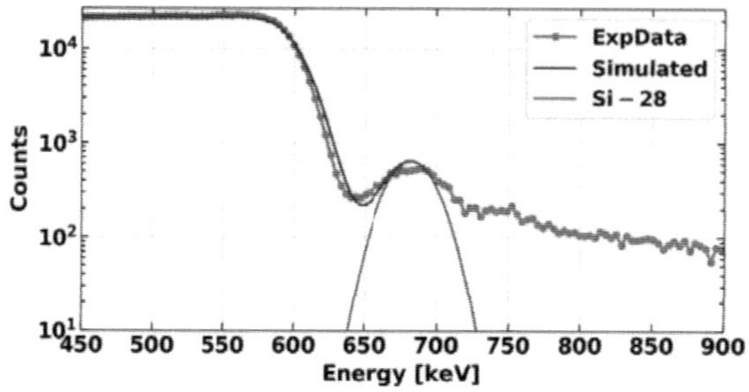

Figure 1. Energy spectrum of backscattered protons for the implanted ^{28}Si at 435 keV in glassy carbon.

The last example is the production of a ^{13}C implanted target. Here we cannot use a carbon matrix, so aluminum was chosen. This choice is particularly well suited for nuclear reaction studies on ^{13}C at low energies since the nuclei in the matrix present a Coulomb Barrier twice as high as that of ^{13}C, strongly suppressing possible contamination of the data with spurious reactions. In turn, the RBS analysis becomes harder to do for the same reason.

Figure 2 shows the RBS analysis spectrum obtained with 1.035 MeV protons (data and SIMNRA simulation) of the ^{13}C implanted target on aluminum. The areal density achieved in this case is $(9.4\pm13)\times10^{16}$ atoms/cm^2 which is also useful for many nuclear reaction studies.

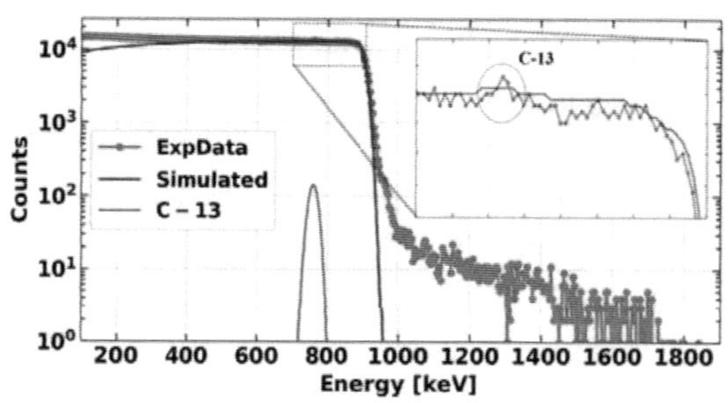

Figure 2. Energy spectrum of backscattered protons for the implanted ^{13}C at 435 keV in aluminum.

Since one of the applications seek in the development of this kind of monoisotopic implanted targets is the study of very low energy nuclear capture cross sections for astrophysics, there is interest of implanting them closer to the surface, that is: at lower energies. This interest added up to the previous one to study molecular contamination in our beams for standard AMS work, and decided to search for molecular beams that survived the argon stripper in the tandem accelerator and that we could identify and transport to the scattering chamber to be implanted at even lower energies. In the next section we describe this effort.

2.2. Molecular beams

A natural graphite cathode is placed in the SNICS, the extracted 35 keV beam of negative ions is transported towards the Electrostatic Spherical Analyser (ESA) that selects particles with a constant E/q value (E=35 keV and the charge state q=-1). The selection of the mass to be injected into the accelerator is made in the following element: the 90° injector magnet. For any given cathode inserted into the SNICS, a mass spectrum (scan) can be produced by slowly changing the electric current in the magnet coils. The field increases proportionally while the beam intensity (current) is measured in the Faraday cup (FC) placed right after the magnet and before the accelerator. Figure 3 shows, in logarithmic scale, the recorded mass spectrum for the graphite cathode. Each peak corresponds to a beam of different mass. This is the well known operation principle of regular mass spectrometers. As can be easily seen, beams made by molecules with several ^{12}C atoms are quite intense.

The peaks amplitude in figure 3 gives a good indication of the relative abundances of the different masses entering the injector magnet.

The highest measured current corresponds to the atomic ^{12}C beam. The diatomic molecular beam is also very intense, although not as much as the atomic one. Beams of molecules with 3, 4, 5 and 6 carbon atoms were seen. The heavier molecular beam useful for our study was the C_6. Beams of larger molecules had much lower intensities and were not well suited for the present study.

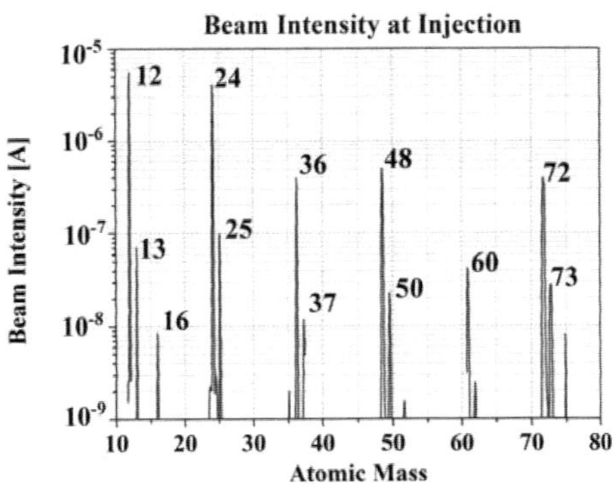

Figure 3. Precise scan of the injector magnet. The vertical axis corresponds the electric current as measured in the FC after the magnet. The horizontal axis is the corresponding mass as calculated from equations (1) and (3). No molecules with 7 carbon atoms or more were seen.

The odd-even effect in the beam intensity is observed in the figure: the abundance of C_4 is larger than that of C_3 and C_6 is also larger than C_5, is worth some qualitative discussion.

Journal of Physics: Conference Series **2619** (2023) 012010 doi:10.1088/1742-6596/2619/1/012010

The formation process of these molecules starts at the heart of the SNICS, where a beam of energetic heavy ions (7 keV cesium atoms) hits the surface of a solid graphite, breaking the bonds that keep the carbon atoms tight together into the graphite. The sputter phenomenon that follows can be thought as similar to what happens when firing a heavy bowling ball to a glass. The dynamical fragmentation of (brittle) solids that follows produces a size distribution which is, classically, a power law or an exponential [47]. Here we witness the interplay of such size distribution with a manifestation of the quantum nature of the molecules: the electron affinity, which has been found to show such an odd-even behavior. A theoretical estimation [48, 49] gives 3.26, 1.58, 3.41, 2.43, and 3.69 eV for C_2, C_3, C_4, C_5 and C_6 respectively. So the fact that an even C_n molecule can hold better the extra electron than the more abundant C_{n-1} molecule, over compensates its abundance.

The mass spectrum in figure 3 is made with a total of 1000 data points, each one taken during 30 s. The injector magnet coils current was changed in 200 mA steps. The coils current "i" is proportional to the magnetic field B in the magnet. The actual relationships for our magnets, injector and high energy, are given in the following equations (1) and (2), respectively [3, 22]:

$$B = 49.0i + 62\ [G] \tag{1}$$

$$B = 35.4i + 48.4\ [G] \tag{2}$$

Since the ESA, located after the SNICS, ensures that all particles entering the injector magnet have constant energy E and charge state q, there is a quadratic dependence between the magnetic field and the mass m:

$$B\rho = 144\sqrt{\frac{mE}{q^2}} \tag{3}$$

where B is the magnetic field in kG and m is the molecule mass in u. The curvature radius is $\rho = 40$ cm (85 cm for the High Energy Magnet).

For this study we have chosen to work with the C_6 molecular beam for two main reasons: its intensity is large enough to be easily handled by our system and for the fact that a molecule with six ^{12}C atoms (72 u) allows different break-up partitions to look for 12, 24, 36, 48, 60 and 72 u at the high energy side of the accelerator.

The energy of single atom ion beams at the exit of a tandem accelerator can be calculated with the following equation:

$$E = eV_i + e(1+q)V_t \tag{4}$$

Where V_i is the total acceleration voltage in the ion source (the SNICS uses a constant value of 35 kV for every beam) and V_t is the terminal voltage, which has been kept down at 200 kV in this study. However, for molecular beams the previous equation needs to be adjusted as follows:

$$E = e\left(\frac{m_r}{m_t}\right)V_i + e\left(\frac{m_r}{m_t} + q\right)V_t\ [keV] \tag{5}$$

where m_r is the molecular mass after the stripper and m_t is the mass of the injected molecule (72 u for C_6), and finally q is the molecular charge state after the stripper.

The beam at the exit of the tandem accelerator is directed to the 90° magnet in the NPL whose field serves as a magnetic rigidity filter to chose a specific beam to be delivered to the reaction chamber.

Figure 4 shows, for a given mass and charge state of the molecules, the energy they should have to be bent exactly 90° as a function of the magnetic field. The black dots in every parabolic curve signals the energy that this particular molecule at that particular charge state should be given by the accelerator, as calculated with equation 5, given that the molecule originally injected in the accelerator is the C_6.

To measure the energy of the incoming particles, a Passivated Implanted Planar (PIPS) detector was placed at 0° in the reaction chamber of the NPL facing the beam. It was energy calibrated using a triple alpha source (^{239}Pu, ^{241}Am and ^{244}Cm) where we obtained:

$$E(C) = 3.223C + 96.92\ [keV] \tag{6}$$

where C is the channel number and E the particle energy.

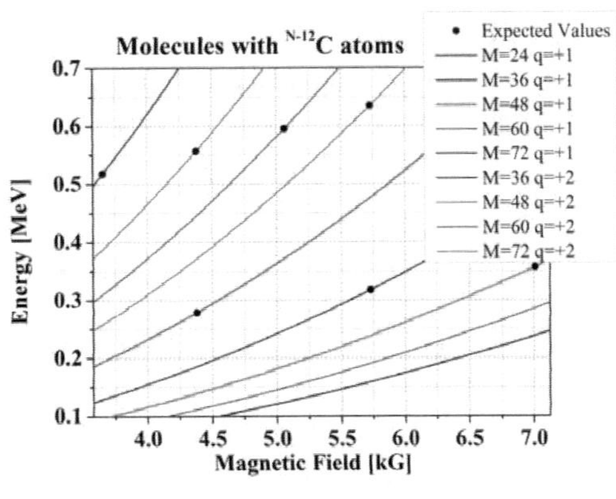

Figure 4. Magnetic field versus Energy for the different molecular fragments at the exit of the accelerator. Each line signals the magnetic rigidity selected for a given mass and charge state.

The identity of the molecular beam reaching this detector in the reaction chamber will be determined by the simultaneous measurement of their energy and magnetic rigidity.

Keeping the accelerator's terminal voltage fixed at 200 kV, we let the accelerated beam into the magnet of the NPL. The magnetic field is then slowly changed by increasing the current in the coils between 100 A and 200 A in steps of 5 A. Keeping the FC in place during the first scan, we made sure that no intense beam is present, to avoid damage to the detector. In a second sweep of the magnetic field the FC was moved out the beam's trajectory and data was taken for 30 seconds in each step.

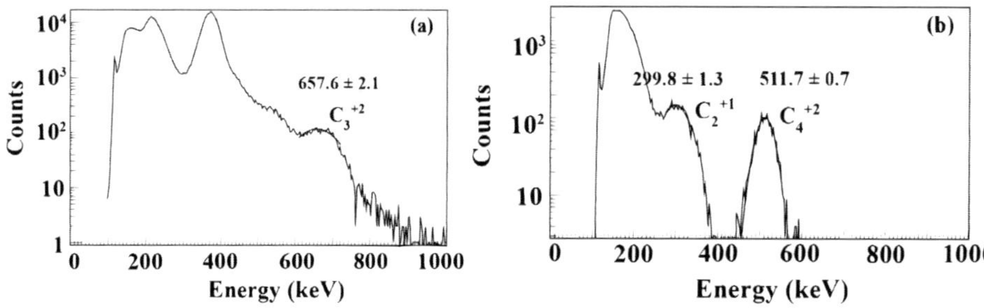

Figure 5. Spectra showing the presence of molecular beams for different magnetic field settings.

3. Results

Not every setting of the magnetic field resulted in an identifiable molecular beam inside the chamber, nor every peak corresponds to an identifiable molecular beam.

Figure 5 shows the spectra a) and b) measured with the PIPS detector at 0° inside the chamber. In all cases, a large flux of particles with energies around and below 200 keV was present. These beams could not be identified. In 5.a) there is a large flux (beam) with energy below 400 keV that is non-identifiable as a C_n molecular beam. The spectrum in figure 5.b) corresponding to 125 A in the magnets coils is a neat example of two molecular beams with different charge states (C_2^+ and C_4^{++}) where their magnetic rigidity and energy combine to produce the same trajectory along the AMS system. Both beams enter the reaction chamber and hit the silicon detector with different energies.

In summary we found four different molecular species that survived as positive ions at the high energy side of our tandem accelerator: C_2^+, C_3^+, C_3^{++} and C_4^{++}. The data analysis was performed using ROOT analysis tools [50].

4. Conclusions

We show the successful production of monoisotopic thin targets by low energy implantation of three different species: ^{28}Si, ^{29}Si and ^{13}C, taking advantage of the isotope separator of the AMS facility in Mexico. We also demonstrated the existence of bound molecular beams in positive charge states when the tandem accelerator potential is set at 200 kV. A successful identification of $C_{2,3}^+$ and $C_{3,4}^{++}$ molecular beams was made by measuring their energy and magnetic rigidity simultaneously. Another interesting finding of this work occurred while choosing the best suited molecule to inject into the accelerator for the molecular beams study: the intensity of the molecular C_n^- flux out of the SNICS is modulated by the electron affinity of each molecular species. The higher the electron affinity, the larger the intensity. In a future work we plan a careful quantification of the intensities of every beam we have seen to increase the tandem's terminal voltage and follow the survival probability of these molecular beams.

Acknowledgments

This work was supported in part by: DGAPAUNAM IG102023, IG101423 and IN112023. Authors are indebted to Sergio Martínez González for his help preparing the graphite cathodes and Dr. María Rodríguez Ceja for her help in the preparation of this manuscript.

References

[1] Solís C, Chávez E, Ortiz M E, Huerta A, Andrade E and Barrios E 2014 *Nuclear Instruments and Methods in Physics Research Section B: Beam Interactions with Materials and Atoms* **331** 233–237
[2] Chávez E, Ortiz M E and Solís C 2018 *Facultad de Ciencias, Universidad Nacional Autónoma de México* **1st Edition**
[3] Reza G, Andrade E, Acosta L, Góngora B, Huerta A, Marín-Lámbarri D J, Mas-Ruiz J, Ortiz M E, Padilla S, Solís C and Chávez E 2019 *Eur. Phys. J. Plus* **134** 590
[4] Tuniz C 2001 *Radiat.Phys. Chem* **61** 317–322
[5] Brown L 1984 *Annu. Rev. Earth Planet. Sci* **12** 39–59
[6] Gómez-Chávez S, Solís C, Gazzola J, Chávez-Lomelí E R, Mondragón M A, Rodríguez-Ceja M and Martínez-Carrillo M A 2017 *Radiocarbon* **59** 545–557
[7] Solís C, Carrillo M A M, Ceja M R, Lomelí E R C and Christen J A 2018 *El Códice Maya de México, Antes Grolier. INAH. México.*
[8] Méndez-García C G, Romero-Guzmán E T, Hernández-Mendoza H, Solís-Rosales C and Chávez-Lomelí E R 2017 *Journal of Radioanalytical Nuclear Chemistry* **314** 1767–1775
[9] Méndez-García C G, Padilla S, Solís C, los Ríos K D, Chávez E, García R, Acosta L and Huerta A 2019 *Journal of Radioanalytical Nuclear Chemistry* **332** 1455–1460
[10] Solís C, Pérez-Andrade G, Rodríguez-Ceja M, Solís-Meza E, Méndez T, Chávez-Lomelí E, Carrillo M A M and Mondragón M A 2017 *Radiocarbon* **59** 1945–1954
[11] Ardelean C F, Becerra-Valdivia L, Pedersen M W, Schwenninger J L, Oviatt C G, Macías-Quintero J I, Arroyo-Cabrales J, Sikora M, Ocampo-Díaz Y Z E, Rubio-Cisneros I I *et al.* 2020 *Nature* **584** 87–92
[12] Chávez-Solís E M, Solís C, Simões N and Mascaró M 2020 *Scientific Reports* **10** 12812
[13] Araujo-Escalona V, Acosta L, Andrade E, Barrón-Palos L, de Lucio O, Favela F, Huerta A, López E, Rita P S, Solís C and Chávez E 2016 *Journal of Physics C.S* **730** 012003
[14] Araujo-Escalona V, Acosta L, Andrade E, Barrón-Palos L, de Lucio O, Favela F, Huerta A, López E, Rita P S, Solís C and Chávez E 2017 *Phys Procedia* **90** 421–428
[15] Reza G, Zunun-Torres A, Padilla S, Mas-Ruiz J, Marín-Lámbarri D, Acosta L, Amador-Valenzuela P, Andrade E, Belmont D, Charón L *et al.* 2020 *The European Physical Journal Plus* **135** 1–11
[16] Acosta L, Araujo-Escalona V, Chávez E, Andrade E, Barrón-Palos L, Favela F, Flores M, García-Ramírez J, Huerta A, deLucio O, Méndez-García C, Ortiz M E, Padilla S, Sánchez-Benítez A M, Santa-Rita P and Solís C 2017 *Eur Phys J W of C* **165** 01001
[17] Marín-Lámbarri D, García-Ramírez J, Sánchez-Zúñiga E, Padilla S, Acosta L, Chávez E, Cruz-Galindo H, Huerta A, Méndez G, Raya-Arredondo R *et al.* 2020 *Physical Review C* **102** 044601
[18] Synal H A, Stocker M and Suter M 2007 *Nuclear Instruments and Methods in Physics Research Section B: Beam Interactions with Materials and Atoms* **259** 7–13

[19] Middleton R 1989 *Department of Physics, University of Pennsylvania*
[20] Vogel J S 2013 *AIP Conference Proceedings* **1515** 89–98
[21] Vogel J S and Giacomo J A 2017 *Radiocarbon* **59** 957–965
[22] HVEE 2018 Operator Manual 1.0 MV Tandetron for AMS
[23] Cockcroft J and Walton E 1932 *Proc. Roy. Soc.* **1** 619–630
[24] Cockcroft J and Walton E 1932 *Proc. Roy. Soc.* **2** 619–630
[25] Gnaser H 1999 *Nuclear Instruments Methods in Physics Research B: Beam Interactions with Materials and Atoms* **149** 38–52
[26] Gnaser H 2001 *Physical Review B* **63** 045415
[27] Gnaser H 2003 *Applied Surface Science* **203-204** 78–81
[28] Bassi D, Buck U and Laine D 1988 *G. Scoles (Ed.). New York: Oxford university press* **1** 499–524
[29] Fluendy M A and Lawley K P 1973 *Chapman and Hall: London; distributed in the USA by Halsted Press, 1973*
[30] Galindo-Uribarri A, H W Lee H W and Chang K H 1985 *Journal of chemical physics* **87** 3685–3693
[31] Weathers D L, McDaniel F D, Matteson S, Duggan J L, Anthony J M and Douglas M A 1991 *Nuclear Instruments and Methods in Physics Research Section B: Beam Interactions with Materials and Atoms* **56** 889–892
[32] Todd A, Lennard W, Magee C and Xia H 2006 *Nuclear Instruments and Methods in Physics Research Section B: Beam Interactions with Materials and Atoms* **242** 572–575
[33] Hebbard D and Vogl J 1960 *Nuclear Physics* **21** 652–675
[34] Dagley P, Haeberli W and Saladin J 1961 *Nuclear Physics* **24** 353–371
[35] Roughton N, Fritts M, Peterson R, Zaidins C and Hansen C 1974 Thick-target measurement of the (p, γ) stellar reaction rates on the nuclides ^{12}C, ^{29}Si, ^{46}Ti, ^{47}Ti, and ^{56}Fe Tech. rep. Regis College, Denver, Colorado 80221, and Nuclear Physics Laboratory
[36] Bromley D, Gove H, Litherland A, Almqvist E and Paul E 1957 *Canadian Journal of Physics* **35** 1042–1056
[37] Baxter A and Hinds S 1973 *Nuclear Physics A* **211** 7–20
[38] Camacho A G, Diaz-Torres A, Gomes P and Lubian J 2015 *Physical Review C* **91** 014607
[39] Fonseca L, Linares R, Zagatto V, Cappuzzello F, Carbone D, Cavallaro M, Agodi C, Lubian J and Oliveira J 2019 *Physical Review C* **100** 014604
[40] Seuthe S, Becker H, Krauss A, Redder A, Rolfs C, Schröder U, Trautvetter H, Wolke K, Wüstenbecker S, Kavanagh R *et al.* 1987 *Nuclear Instruments and Methods in Physics Research Section A: Accelerators, Spectrometers, Detectors and Associated Equipment* **260** 33–42
[41] Brown T, Deryckx K, García A, Sallaska A, Snover K, Storm D and Wrede C 2009 *Nuclear Instruments and Methods in Physics Research Section B: Beam Interactions with Materials and Atoms* **267** 3302–3308
[42] Lennard W, Setoodehnia K, Chen A and Hendriks J 2011 *Nuclear Instruments and Methods in Physics Research Section B: Beam Interactions with Materials and Atoms* **269** 2726–2729
[43] Wang X, Zhang Y W, Han D, Zhao Y B, Zhao Z Q and Zhang M 2018 *Nuclear Science and Techniques* **29** 1–7
[44] Miranda J, Serrano D, Pineda J, Marín-Lámbarri D, Acosta L, Mendoza-Flores J, Reynoso-Cruces S and Chávez E 2020 *Nuclear Instruments and Methods in Physics Research Section B: Beam Interactions with Materials and Atoms* **477** 23–26
[45] Favela F, Acosta L, Andrade E, Araujo V, Huerta A, de Lucio O, Murillo G, Ortiz M, Policroniades R, Santa Rita P *et al.* 2015 *Physical Review Special Topics-Accelerators and Beams* **18** 123502
[46] Mayer M 1999 *AIP Conference Proceedings* **475** 541–544
[47] Åström J A, Linna R P, Timonen J, Möller P F and Oddershede L 2004 *Physical Review E* **70** 026104
[48] Adamowicz L 1991 *The Journal of Chemical Physics* **94** 1241–1246
[49] Lu S 2004 *The Journal of Chemical Physics* **121** 10495
[50] Brun R and Rademakers F 1997 *Nuclear Instruments and Methods in Physics Research Section A* **389** 81–86

Journal of Physics: Conference Series **2619** (2023) 012011 doi:10.1088/1742-6596/2619/1/012011

Heavy $\Xi_{c/b}$ and $\Xi'_{c/b}$ baryons in the quark model

E. Ortiz-Pacheco
Jozef Stefan Institute, Jamova 39, 1000 Ljubljana, Slovenia
E-mail: emmanuel.ortiz@ijs.si

R. Bijker
Instituto de Ciencias Nucleares, Universidad Nacional Autónoma de México, Coyocacán, 04510 Ciudad de México, México
E-mail: bijker@nucleares.unam.mx

Abstract. We present a study of the low-lying $\Xi_{c/b}$ and $\Xi'_{c/b}$ baryon states based on masses, strong and radiative decays. In a harmonic oscillator quark model analysis it is found that the majority of the detected states can be interpreted as S- and P-wave excitations.

1. Introduction

According to the Particle Data Group compilation [1], there are 15 single-charm cascade baryons, namely $\Xi_{c/b}$ and $\Xi'_{c/b}$, whose angular momentum and parity for certain states have yet to be determined. The LHCb collaboration has recently observed three negative parity states of Ξ_c^0, specifically $\Xi_c^0(2923)$, $\Xi_c^0(2939)$, and $\Xi_c^0(2965)$, which follow a mass rule of roughly 126 MeV with respect to the excited negative parity Ω_c^0 states. Therefore, it is suggested that these three new negative parity excited Ξ_c^0 states belong to the same flavor $SU_{\rm f}(3)$ multiplet as the negative parity excited Ω_c^0 states, a flavor sextet **6**. Additionally, it is assumed that $\Xi_c(2965)$ is the same state as the previously observed $\Xi_c(2970)$.

This observation made by the LHCb collaboration is important as it sheds light on the fundamental binding mechanism of matter at a basic level and emphasizes the significance of comprehending the interactions between quarks. The quantum numbers of the states are assigned using the quark model systematics, and the equal spacing mass rule for the $SU_{\rm f}(3)$ ground states in the Gell-Mann-Okubo and Gürsey-Radicati mass formulas is similar to the spacing of the three new negative parity excited Ξ_c^0 states. Nonetheless, further research is required to validate these findings and gain a more in-depth understanding of the quark structure of hadrons.

2. $\Xi_{c/b}$ and $\Xi'_{c/b}$ baryons

The $\Xi_{c/b}$ and $\Xi'_{c/b}$ baryons can be described in the quark model as qqQ configurations, two light quarks and one heavy quark. The light quarks q correspond to the u, d or s flavors, whereas the heavy quark Q can be either c or b. The masses of the singly-heavy cascade baryons are calculated in a harmonic oscillator quark model with spin, spin-orbit, isospin, and flavor

Table 1. Classification of highest charge state of sextet $\Xi'_{c/b}$ baryons (top) and anti-triplet $\Xi_{c/b}$ baryons (bottom).

State	Wave function	(n_ρ, n_λ)	L^P	J^P
$^2\Xi'_Q$	$\frac{1}{\sqrt{2}}(us+su)Q\,[\psi_0\chi_\lambda]$	$(0,0)$	0^+	$\frac{1}{2}^+$
$^4\Xi'_Q$	$\frac{1}{\sqrt{2}}(us+su)Q\,[\psi_0\chi_S]$	$(0,0)$	0^+	$\frac{3}{2}^+$
$^2\lambda(\Xi'_Q)_J$	$\frac{1}{\sqrt{2}}(us+su)Q\,[\psi_\lambda\chi_\lambda]_J$	$(0,1)$	1^-	$\frac{1}{2}^-, \frac{3}{2}^-$
$^4\lambda(\Xi'_Q)_J$	$\frac{1}{\sqrt{2}}(us+su)Q\,[\psi_\lambda\chi_S]_J$	$(0,1)$	1^-	$\frac{1}{2}^-, \frac{3}{2}^-, \frac{5}{2}^-$
$^2\rho(\Xi'_Q)_J$	$\frac{1}{\sqrt{2}}(us+su)Q\,[\psi_\rho\chi_\rho]_J$	$(1,0)$	1^-	$\frac{1}{2}^-, \frac{3}{2}^-$
$^2\Xi_Q$	$\frac{1}{\sqrt{2}}(us-su)Q\,[\psi_0\chi_\rho]$	$(0,0)$	0^+	$\frac{1}{2}^+$
$^2\lambda(\Xi_Q)_J$	$\frac{1}{\sqrt{2}}(us-su)Q\,[\psi_\lambda\chi_\rho]_J$	$(0,1)$	1^-	$\frac{1}{2}^-, \frac{3}{2}^-$
$^2\rho(\Xi_Q)_J$	$\frac{1}{\sqrt{2}}(us-su)Q\,[\psi_\rho\chi_\lambda]_J$	$(1,0)$	1^-	$\frac{1}{2}^-, \frac{3}{2}^-$
$^4\rho(\Xi_Q)_J$	$\frac{1}{\sqrt{2}}(us-su)Q\,[\psi_\rho\chi_S]_J$	$(1,0)$	1^-	$\frac{1}{2}^-, \frac{3}{2}^-, \frac{5}{2}^-$

interactions as follows

$$H = H_{\rm ho} + A\,\vec{S}\cdot\vec{S} + B\,\vec{L}\cdot\vec{S} + E\,\vec{I}\cdot\vec{I} + G\,C_{2SU_{\rm f}(3)}\ , \tag{1}$$

where the Hamiltonian of the harmonic oscillator for this system [2]

$$\begin{aligned} H_{\rm ho} &= \sum_i \left(m_i + \frac{p_i^2}{2m_i}\right) + \frac{1}{2}C\sum_{i<j}(\vec{r}_i - \vec{r}_j)^2 \\ &= M + \frac{P^2}{2M} + \frac{p_\rho^2}{2m_\rho} + \frac{p_\lambda^2}{2m_\lambda} + \frac{1}{2}m_\rho\omega_\rho^2\rho^2 + \frac{1}{2}m_\lambda\omega_\lambda^2\lambda^2\ , \end{aligned} \tag{2}$$

is expressed in terms of the Jacobi coordinates

$$\begin{aligned} \vec{\rho} &= (\vec{r}_1 - \vec{r}_2)/\sqrt{2}\ , \\ \vec{\lambda} &= (\vec{r}_1 + \vec{r}_2 - 2\vec{r}_3)/\sqrt{6}\ , \\ \vec{R} &= \frac{m_q(\vec{r}_1 + \vec{r}_2) + m_Q\vec{r}_3}{2m_q + m_Q}\ , \end{aligned} \tag{3}$$

and their conjugate momenta. The reduced masses are given by $m_\rho = (m_{u/d} + m_s)/2$, and $m_\lambda = 3m_\rho m_Q/M$, with $M = 2m_\rho + m_Q$, and the oscillator frequencies by $\omega_\rho = \sqrt{3C/m_\rho}$ and $\omega_\lambda = \sqrt{3C/m_\lambda}$. The baryons $\Xi_{c/b}$ and $\Xi'_{c/b}$ are isospin doublets with $I = 1/2$. Since both the isospin I and the sum of the quark masses M have the same value for all baryons, they are absorbed into M_0. The mass formula is given by

$$\begin{aligned} E &= M_0(^2\Xi_Q) + \hbar\omega_\rho n_\rho + \hbar\omega_\lambda n_\lambda + A\left[S(S+1) - \frac{3}{4}\right] + B\frac{1}{2}\left[J(J+1) - L(L+1) - S(S+1)\right] \\ &\quad + G\frac{1}{3}\left[p(p+3) + q(q+3) + pq - 4\right]\ , \end{aligned} \tag{4}$$

where S, L, and J are the spin, orbital angular momentum and total angular momentum, respectively.

We only consider baryons associated with the ground-state configuration ψ_0 with $(n_\rho, n_\lambda) = (0,0)$, and with one quantum of excitation, either in the λ mode, ψ_λ, with $(n_\rho, n_\lambda) = (0,1)$, or in the ρ mode, ψ_ρ, with $(n_\rho, n_\lambda) = (1,0)$. A summary of the baryon wave functions together with their quantum numbers is presented in Table 1.

3. Masses and decay widths

The values of the parameters in the mass formula of Eq. (1) were determined by studying some well established singly-heavy baryons [3][4]. In Figs. 1 and 2 we present the mass spectra of the $\Xi_{c/b}$ and the $\Xi'_{c/b}$ baryons, respectively. We compare our results (in red points) with the available experimental data (in black and blue lines) [1]. We find a good agreement between the experiment and our calculated theoretical masses for the majority of the $\Xi_{c/b}$ and $\Xi'_{c/b}$ baryons. The explicit assignments between the data and our states can be seen in Tables 2 and 3 [5].

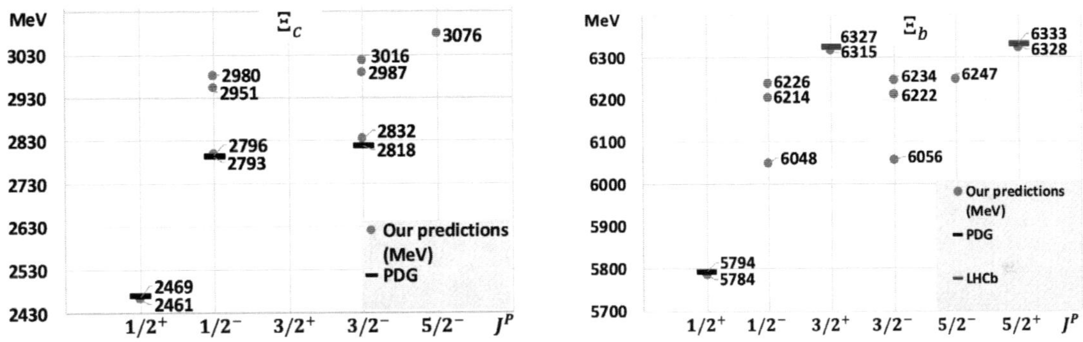

Figure 1. The mass spectra of Ξ_c (left) and Ξ_b (right) along with tentative quantum number assignments, are shown. The theoretical predictions (represented by red dots) are compared with experimental results obtained by the LHCb collaboration [6] [7] (depicted as blue lines), as well as with the Particle Data Group compilation (depicted as black lines) [1].

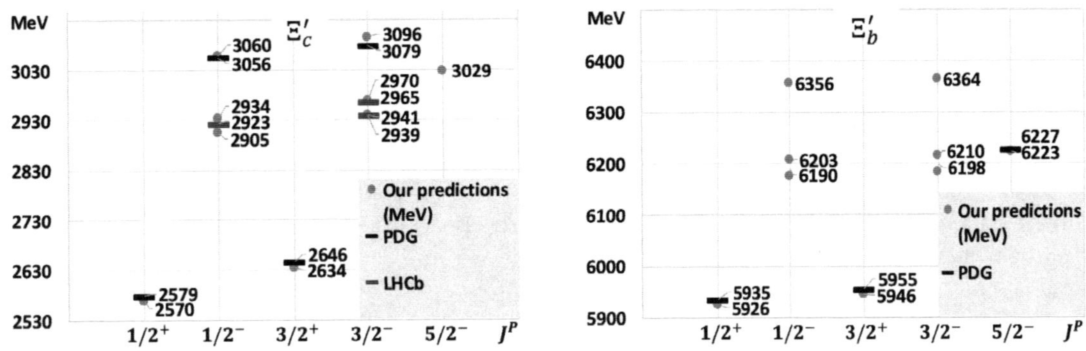

Figure 2. The mass spectra of Ξ'_c (left) and Ξ'_b (right) along with tentative quantum number assignments, are shown. The theoretical predictions (represented by red dots) are compared with experimental results obtained by the LHCb collaboration [6] [7] (depicted as blue lines), as well as with the Particle Data Group compilation (depicted as black lines) [1].

Table 2. Comparison between the strong total decay widths for charmed baryons Ξ_c and $\Xi_c^{'}$ calculated within EEM and $^3P_0$ model with the chiral quark model [8] and the available experimental data. All values are expressed in MeV.

State	This work EEM	$^3P_0$	χQM [8]	Exp [1]	Baryon
$^4\Xi_c'$	1.23	0.02		2.14 ± 0.19	$\Xi_c(2645)^+$
	1.23	0.02		2.35 ± 0.22	$\Xi_c(2645)^0$
$^2\lambda(\Xi_c')_{1/2^-}$	1.75	0.79	21.67		
$^4\lambda(\Xi_c')_{1/2^-}$	4.83	0.53	37.05	7.1 ± 2.0	$\Xi_c(2923)^0$
$^2\lambda(\Xi_c')_{3/2^-}$	11.53	3.08	20.89	14.8 ± 9.1	$\Xi_c(2930)^+$
	11.53	3.08		10.2 ± 1.4	$\Xi_c(2939)^0$
$^4\lambda(\Xi_c')_{3/2^-}$	3.88	2.04	12.33	14.1 ± 1.6	$\Xi_c(2965)^0$
$^4\lambda(\Xi_c')_{5/2^-}$	29.27	5.43	20.20		
$^2\rho(\Xi_c')_{1/2^-}$	10.73	6.26		7.8 ± 1.9	$\Xi_c(3055)^+$
$^2\rho(\Xi_c')_{3/2^-}$	25.39	3.70		3.6 ± 1.1	$\Xi_c(3080)^+$
	25.39	3.70		5.6 ± 2.2	$\Xi_c(3080)^0$
$^2\lambda(\Xi_c)_{1/2^-}$	0.01	0.41	3.61	8.9 ± 1.0	$\Xi_c(2790)^+$
	0.01	0.41		10.0 ± 1.1	$\Xi_c(2790)^0$
$^2\lambda(\Xi_c)_{3/2^-}$	0.28	0.54	2.11	2.43 ± 0.26	$\Xi_c(2815)^+$
	0.28	0.54		2.54 ± 0.25	$\Xi_c(2815)^0$
$^2\rho(\Xi_c)_{1/2^-}$	4.57	0.70			
$^4\rho(\Xi_c)_{1/2^-}$	10.20	0.45			
$^2\rho(\Xi_c)_{3/2^-}$	19.45	3.76			
$^4\rho(\Xi_c)_{3/2^-}$	7.60	2.51			
$^4\rho(\Xi_c)_{5/2^-}$	44.91	4.55			

In the same tables we show the results for the strong decays in two models: the elementary emission model (EEM) and the $^3P_0$ model. The values correspond to the total strong decay widths. In general we find a reasonable agreement between our calculation of the strong decay widths with the experimental data.

Finally, we present our results of the electromagnetic decay widths for the S- and P-wave single charm baryons. In Table 4 we present a comparison with the experimental data on radiative decay widths of the $\Xi_c(2790)$ and $\Xi_c(2915)$ baryons [9]. The Belle collaboration found that the decay width for the neutral baryons is much larger than that for the charged baryons (albeit with large error bars), in agreement with the present calculations as well as with the results of the chiral quark model [8], thus confirming the assignment of these statas as a λ-mode excitation. In Tables 5 and 6 we show the electromagnetic decay widths for S- and P-wave single-heavy cascade baryons. The results are in qualitative agreement with those obtained in the chiral quark model.

4. Summary and conclusions

We constructed the low-lying baryon states $\Xi_{c/b}$ and $\Xi'_{c/b}$ in the quark model for S- and P-wave and make an analysis related with their quantum numbers and the available experimental data. The assignment of quantum numbers to the experimentally observed single-charm and single-

Table 3. As in Tab. 2 but for Ξ_b and $\Xi_b^{'}$.

State	This work EEM	$^3P_0$	χQM [8]	Exp [1]	Baryon
$^2\Xi_b^{'}$	–	–	0.08	< 0.08	
$^4\Xi_b^{'}$	0.48	0.02	0.98	0.90 ± 0.18	$\Xi_b(5945)^0$
	0.48	0.02		1.65 ± 0.33	$\Xi_b(5955)^-$
$^2\lambda(\Xi_b^{'})_{1/2^-}$	1.78	0.86	27.05		
$^4\lambda(\Xi_b^{'})_{1/2^-}$	3.99	0.65	32.24		
$^2\lambda(\Xi_b^{'})_{3/2^-}$	8.30	2.92	24.15		
$^4\lambda(\Xi_b^{'})_{3/2^-}$	2.33	1.83	15.83		
$^4\lambda(\Xi_b^{'})_{5/2^-}$	13.80	3.36	24.39	$18.6^{+5.2}_{-4.3}$	$\Xi_b(6227)^0$
	13.80	3.36		19.9 ± 2.6	$\Xi_b(6227)^-$
$^2\rho(\Xi_b^{'})_{1/2^-}$	10.93	5.88			
$^2\rho(\Xi_b^{'})_{3/2^-}$	13.83	3.08			
$^2\lambda(\Xi_b)_{1/2^-}$	–	–	2.88		
$^2\lambda(\Xi_b)_{3/2^-}$	–	–	2.95		
$^2\rho(\Xi_b)_{1/2^-}$	1.94	0.55			
$^4\rho(\Xi_b)_{1/2^-}$	4.21	0.36			
$^2\rho(\Xi_b)_{3/2^-}$	8.44	1.90			
$^4\rho(\Xi_b)_{3/2^-}$	2.47	1.90			
$^4\rho(\Xi_b)_{5/2^-}$	13.45	2.16			

Table 4. Radiative decay widths of $\Xi_c(2790)$ and $\Xi_c(2815)$ baryons in keV.

Decay	Our work	χQM [8]	Exp [9]
$\Xi_c(2790)^+ \to {}^2\Xi_c^+ + \gamma$	5.4	4.6	< 350
$\Xi_c(2790)^0 \to {}^2\Xi_c^0 + \gamma$	239.3	263.0	800 ± 320
$\Xi_c(2815)^+ \to {}^2\Xi_c^+ + \gamma$	2.4	2.8	< 80
$\Xi_c(2815)^0 \to {}^2\Xi_c^0 + \gamma$	344.6	292.0	$320 \pm 45^{+45}_{-80}$

bottom baryons was based on a combination of energy systematics, strong and electromagnetic decay widths. Overall, a reasonable agreement was found with the available experimental data.

Acknowledgments
It is a pleasure to thank Elena Santopinto, Alessandro Giachino and Hugo García-Tecocoatzi for interesting discussions. This work was supported by grant No. IG101423 from PAPIIT, DGAPA-UNAM, Mexico, and by grant No. 251817 from CONACyT, Mexico. EOP acknowledges the projects (P1-0035) and (J1-3034) with financial support from the Slovenian Research Agency.

Table 5. Radiative decay widths of S- and P-wave Ξ_c' and Ξ_c baryons in keV.

	$^2\Xi_c' + \gamma$		$^4\Xi_c' + \gamma$		$^2\Xi_c + \gamma$		
	+	0	+	0	+	0	
$^2\Xi_c'$					15.1	0.3	
					42.3	0.0	[8]
$^4\Xi_c'$	0.0	0.9			59.8	1.3	
	0.0	3.0			139	0.0	[8]
$^2\lambda(\Xi_c')_{1/2^-}$	0.4	183.5	0.4	0.2	42.9	0.9	
	0.0	472.0	1.6	1.0	46.4	0.0	[8]
$^2\lambda(\Xi_c')_{3/2^-}$	17.0	401.7	0.7	0.3	57.5	1.2	
	12.1	302.0	1.6	1.0	46.1	0.0	[8]
$^4\lambda(\Xi_c')_{1/2^-}$	1.5	0.7	0.3	20.3	27.2	0.6	
	0.3	0.2	0.2	125.0	14.5	0.0	[8]
$^4\lambda(\Xi_c')_{3/2^-}$	6.5	2.8	0.5	122.4	99.4	2.1	
	2.1	1.2	1.6	187.0	54.6	0.0	[8]
$^4\lambda(\Xi_c')_{5/2^-}$	7.3	2.9	12.0	293.2	92.9	2.0	
	1.6	0.9	2.3	192.0	32.0	0.0	[8]
$^2\rho(\Xi_c')_{1/2^-}$	16.3	26.6	4.7	7.7	731.2	15.5	
$^2\rho(\Xi_c')_{3/2^-}$	20.9	34.3	6.4	10.5	783.4	16.6	
$^2\lambda(\Xi_c)_{1/2^-}$	2.3	0.0	0.2	0.0	5.4	239.3	
	1.4	0.0	0.4	0.0	4.6	263.0	[8]
$^2\lambda(\Xi_c)_{3/2^-}$	4.6	0.1	0.6	0.0	2.4	344.6	
	2.3	0.0	1.0	0.0	2.8	292.0	[8]
$^2\rho(\Xi_c)_{1/2^-}$	157.2	3.3	1.8	0.0	16.0	26.2	
	128.0	0.0	0.2	0.0	1.4	5.6	[8]
$^2\rho(\Xi_c)_{3/2^-}$	585.1	12.4	2.8	0.1	20.6	33.7	
	110.0	0.0	0.5	0.0	1.9	7.5	[8]
$^4\rho(\Xi_c)_{1/2^-}$	5.4	0.1	12.5	0.3	9.8	16.1	
	0.4	0.0	43.4	0.0	0.7	3.0	[8]
$^4\rho(\Xi_c)_{3/2^-}$	21.2	0.4	122.4	2.6	34.6	56.6	
	1.8	0.0	58.1	0.0	2.8	11.2	[8]
$^4\rho(\Xi_c)_{5/2^-}$	21.7	0.5	445.5	9.5	30.8	50.5	

Table 6. Radiative decay widths of S- and P-wave Ξ_b' and Ξ_b baryons in keV.

	$^2\Xi_b' + \gamma$		$^4\Xi_b' + \gamma$		$^2\Xi_b + \gamma$		
	0	−	0	−	0	−	
$^2\Xi_b'$					33.2	0.7	
					84.6	0.0	[8]
$^4\Xi_b'$	0.0	0.0			49.5	1.0	
	5.2	15.0			104	0.0	[8]
$^2\lambda(\Xi_b')_{1/2^-}$	48.2	46.4	0.3	0.5	53.7	1.1	
	76.3	190.0	0.9	3.5	72.2	0.0	[8]
$^2\lambda(\Xi_b')_{3/2^-}$	101.2	116.0	0.3	0.5	57.9	1.2	
	43.9	92.3	0.9	3.6	72.8	0.0	[8]
$^4\lambda(\Xi_b')_{1/2^-}$	0.5	0.8	6.5	5.3	30.3	0.6	
	0.3	1.5	69.5	164.0	34.0	0.0	[8]
$^4\lambda(\Xi_b')_{3/2^-}$	1.4	2.5	35.6	35.4	90.2	1.9	
	0.7	2.0	47.5	104.0	94.0	0.0	[8]
$^4\lambda(\Xi_b')_{5/2^-}$	1.1	1.9	64.1	76.1	64.8	1.4	
	0.4	1.9	41.5	88.2	47.7	0.0	[8]
$^2\rho(\Xi_b')_{1/2^-}$	12.3	20.2	5.2	8.6	698.5	14.8	
$^2\rho(\Xi_b')_{3/2^-}$	13.2	21.6	5.6	9.2	705.7	15.0	
$^2\lambda(\Xi_b)_{1/2^-}$	0.2	0.0	0.0	0.0	72.8	80.0	
	1.3	0.0	2.0	0.0	63.6	135.0	[8]
$^2\lambda(\Xi_b)_{3/2^-}$	0.3	0.0	0.1	0.0	79.3	85.6	
	1.7	0.0	2.6	0.0	68.3	147.0	[8]
$^2\rho(\Xi_b)_{1/2^-}$	120.9	2.6	1.2	0.0	11.9	19.5	
	94.3	0.0	0.6	0.0	1.9	7.2	[8]
$^2\rho(\Xi_b)_{3/2^-}$	296.9	6.3	1.3	0.0	12.9	21.1	
	69.4	0.0	0.8	0.0	2.1	8.1	[8]
$^4\rho(\Xi_b)_{1/2^-}$	1.9	0.0	13.7	0.3	6.7	10.9	
	0.2	0.0	80.0	0.0	0.9	3.6	[8]
$^4\rho(\Xi_b)_{3/2^-}$	5.9	0.1	91.5	1.9	19.9	32.5	
	0.8	0.0	78.0	0.0	2.9	11.4	[8]
$^4\rho(\Xi_b)_{5/2^-}$	4.6	0.1	193.9	4.1	14.1	23.1	

References

[1] Zyla P *et al.* (Particle Data Group) 2020 *PTEP* **2020** 083C01

[2] Isgur N and Karl G 1978 *Phys. Rev. D* **18**(11) 4187–4205

[3] Santopinto E, Giachino A, Ferretti J, García-Tecocoatzi H, Bedolla M A, Bijker R and Ortiz-Pacheco E 2019 *The European Physical Journal C* **79** 1012

[4] Ortiz-Pacheco E, Bijker R, Giachino A and Santopinto E 2020 *J. Phys. Conf. Ser.* **1610** 012011 (*Preprint* 2004.09409)

[5] Bijker R, García-Tecocoatzi H, Giachino A, Ortiz-Pacheco E and Santopinto E 2022 *Phys. Rev. D* **105**(7) 074029

[6] Aaij R *et al.* (LHCb Collaboration) 2020 *Phys. Rev. Lett.* **124**(22) 222001

[7] Aaij R *et al.* (LHCb Collaboration) 2022 *Phys. Rev. Lett.* **128**(16) 162001

[8] Wang K L, Yao Y X, Zhong X H and Zhao Q 2017 *Phys. Rev. D* **96**(11) 116016

[9] Yelton J *et al.* (The Belle Collaboration) 2020 *Phys. Rev. D* **102**(7) 071103

Journal of Physics: Conference Series **2619** (2023) 012012 doi:10.1088/1742-6596/2619/1/012012

Investigating the Primordial Universe through nuclear physics

R.G. Pizzone[(a,b)], C. Spampinato[(a)], M.L. Sergi[(a,b)], R. Spartá[(a,c)], T. Zanatta-Martinez[(a,b)] for the ASFIN collaboration

[(a)]INFN Laboratori Nazionali del Sud, Catania, Italy
[(b)]Dipartimento di Fisica e Astronomia, Universitá di Catania, Catania, Italy
[(c)]Universitá Kore di Enna, Enna, Italy

E-mail: rgpizzone@lns.infn.it

Abstract. Big Bang Nucleosynthesis (BBN) requires several nuclear physics inputs and nuclear reaction rates. An up-to-date compilation of direct cross sections of is given, being these ones among the most uncertain bare-nucleus cross sections. A particular attention is devoted to recently indirectly measured cross-section which give important hints for the nuclear astrophysics community.

In reality, a significant experimental effort has been made over the past 10 years to explore reactions important to the BBN and determine their astrophysical S(E)-factor using the Trojan Horse Method (THM). Then, numerical calculations are made in the relevant temperature ranges for BBN ($0.01<T_9<10$) to determine the reaction rates and the relative error for the four reactions of interest. The effects of these values on the calculated primordial abundances and isotopical composition for H, He, and Li were then assessed by using them as input physics for computations of primordial nucleosynthesis. Additionally, recent findings regarding the 7Be(n,$alpha$)4He reaction rate were taken into consideration.These were put up against estimates of primordial abundance derived from observation at various astrophysical places. Additionally, perspectives on reactions will be examined.

1. Introduction

BBN is one of the pillars of the Big Bang theory, along with the Hubble expansion and the Cosmic Microwave Background (CMB) radiation [1]. BBN explores the early universe, often known as the radiation dominated era, from a fraction of a second to several minutes. It naturally plays a significant part in establishing the link between cosmology and nuclear physics because it involves events that took place at temperatures below 1 MeV. The Standard Big Bang Nucleosynthesis model (SBBN), which focuses solely on the products of the BBN, predicts that only the production of light nuclei (^{2}H,^{3,4}He,^{7}Li) in measurable quantities, beginning with protons and neutrons, will occur. The abundances of these isotopes in the proper astrophysical conditions today, with the exception of ^{3}He and lithium, are largely compatible with SBBN predictions [3]. The baryon-to-photon ratio, η, which is the only free parameter in the currently accepted model of the SBBN, is constrained by comparing the primordial abundances inferred from precise WMAP CMB observations with those estimated abundances. The value we use in our computations is $\eta = 6.16 \pm 0.15 \times 10^{-10}$, which is based on a recent observation [4]. Nuclear reaction rates are one of many nuclear physics inputs needed for BBN nucleosynthesis, and they

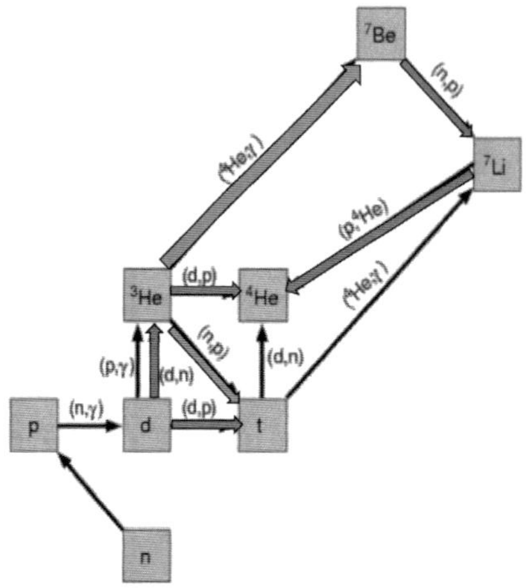

Figure 1. Standard Big Bang nucleosynthesis network. The reactions recently (2020-2022) measured by the ASFIN group are marked by green arrows by means of THM or ANC method. The red ones represent cross sections measured by the same group in previous years.

are crucial. Only 12 reactions significantly contribute to the BBN nuclear reaction network due to the very small number of important nuclear species involved [5]. A number of those involve neutrons and radioactive ions and are currently not known with sufficient precision (see table 1).

$n \leftrightarrow p$ (1)	$p(n,\gamma)d$ (2)	$d(p,\gamma)^3He$ (3)	$d(d,p)t^{(\dagger)}$ (4)
$d(d,n)^3He^{(\dagger)}$ (5)	$^3He(n,p)t$ (6)	$t(d,n)^4He$ (7)	$^3He(d,p)^4He^{(\dagger)}$ (8)
$^3He(\alpha,\gamma)^7Be$ (9)	$t(\alpha,\gamma)^7Li$ (10)	$^7Be(n,p)^7Li$ or $^7Be(n,\alpha)^4He^{(11)(\dagger)}$	$^7Li(p,\alpha)^4He^{(\dagger)}$ (12)

Table 1. Nuclear reactions of greatest relevance for BBN. The reactions whose cross sections were already measured with the Trojan Horse method are marked with a † symbol.

The reaction rates are determined using the low-energy cross sections that are now accessible for processes that are also a crucial input for a number of other unresolved astronomical issues, such as the so-called "lithium depletion" in the Sun or other galactic stars [6]. Cross sections of the order of a few hundred keV should be measured in the Gamow window [7]. These reactions have received extensive research over the past few decades, with particular emphasis on direct observations made at the appropriate astronomical energy, sometimes in subterranean labs [8]. However, there are few direct data at such low energies for many of the pertinent processes (mainly due to challenges associated with the existence of the Coulomb barrier) and the reaction cross-section within the Gamow window extrapolated from direct measurement at high energies.

2. The Trojan Horse Method

Alternative methods for determining bare nucleus cross sections of astrophysical interest are needed. In this context a number of indirect methods, e.g. the Coulomb dissociation (CD) [9], the Asymptotic Normalization Coefficient method (ANC)[10] and references therein and the Trojan-horse method (THM) were developed [11]. For further information on the development and first principles of the method please refer to [12]. The latter has already been applied several times to reactions connected with fundamental astrophysical problems such as primordial nucleosynthesis (as sketched in figure 1)[14, 15], lithium problem [17, 25], light elements depletion [23], AGB [26, 27] and Novae nucleosynthesis [28]. It was also applied to reactions induced by radioactive ion beams [29, 30] and neutrons [31]. THM selects the quasi-free (QF) contribution of an appropriate three-body reaction performed at energies well above the Coulomb barrier to extract a charged particle two-body cross section at energies of astrophysical interest. The idea of the THM [11] is to extract the cross section of an astrophysically relevant two-body reaction

$$A + x \rightarrow c + C \tag{1}$$

at low energies from a suitable chosen three-body quasi-free reaction

$$A + a \rightarrow c + C + S \tag{2}$$

In this approach S acts as a spectator to the $A + x \rightarrow c + C$ binary interaction.This is done with the help of direct processes theory assuming that the Trojan Horse nucleus a has a strong $x \oplus S$ cluster structure [34, 35]. In many applications, this assumption is trivially fulfilled e.g. a = deuteron, x = proton, S= neutron. If the bombarding energy E_A is chosen high enough to overcome the Coulomb barrier in the entrance channel of the three-body reaction, both Coulomb barrier and electron screening effects are negligible. The polar approximation, used in the standard THM prescription has been extensively verified [32] and constitutes a powerful validity test for the method which strenghtens the theoretical approach. We refer to [19, 20] for further and advanced theoretical approach to the method.

We just underline that THM allows to link the three-body cross section which is measured in the laboratory with the half-off-energy shell cross section of the binary process of astrophysical interest. Then after inclusion of the Coulombian effects data are then compared and normalized to direct data, at the higher energies available. After that phase, the reaction rate is calculated according to the standard prescriptions. A summary of results presented to the nuclear astrophysics community by means of THM is reported in table 2.

Several of the reactions that are important for the SBBN, such as ^{7}Li(p,α)^{4}He, ^{2}H(d,p)^{3}H, ^{2}H(d,n)He and ^{3}He (d, p)^{4}He, ^{7}Be(n,α)^{4}He were investigated using the THM in the relevant energy range during an experimental effort that took place over the past ten years [36, 13, 18, 37, 23, 38].

Rates have been generated for the four reactions indicated above (from a composition of direct and THM data, as published in [15]. This is the standard practice used in earlier works (see, for example, [39, 40, 41]). We have properly accounted for the experimental measurement errors for the 4 reactions of interest, allowing us to assess the corresponding uncertainties in the reaction rates. Most recently the ^{7}Be(n,α)^{4}He and ^{7}Be(n,p)^{7}Li have been measured by means of THM and data are reported elsewhere [45, 46]. Even in those cases no real step aiming at the cosmological lithium problem was made, thus suggesting that its solution lies elsewhere.

2.1. The ^{3}He(n,p)^{3}H reaction

Most recently the attention was focused on the ^{3}He(n,p)^{3}H reaction, taking advantage of the application of THM to neutron induced reactions.

One of the most important neutron-induced reactions in BBN is the ^{3}He(n,p)^{3}H process, which has a significant effect on the generation of the primordial isotopes of He and Li. The reaction rate is defined by the ^{3}He(n,p)^{3}H cross section in the energy range $0 \leq E_{cm} \leq 0.4$ MeV at the temperatures important for forecasting Big Bang yields. Coon et al. [47] conducted the initial research on this reaction in 1950 in the $0.1 \leq E_{cm} \leq 30$ MeV range using a neutron beam. Errors ended up being about 30%.

Table 2. BBN two-body reactions recently studied via Trojan Horse Method at the energies of interest together with the relative bibliographical references.

Binary reaction	Indirect reaction	TH nucleus	Reference
^{7}Li(p,α)^{4}He	^{7}Li(d,α α)n	d = (p$\oplus$n)	[18]
^{6}Li(d,α)^{4}He	^{6}Li(^{6}Li,α α)^{4}H	^{6}Li = ($\alpha\oplus$d)	[16]
^{6}Li(p,α)^{3}He	^{6}Li(d,α ^{3}He)n	d = (p$\oplus$n)	[25]
^{2}H(^{3}He,p)^{4}He	^{6}Li(^{3}He,p α)^{4}He	^{3}He = (p$\oplus$n)	[43]
^{2}H(d,p)^{3}H	^{2}H(^{6}Li,t p)^{4}He	^{6}Li = ($\alpha\oplus$d)	[13]
^{3}He(d,p)^{4}He	^{3}He(^{6}Li,p α)^{4}He	^{6}Li = ($\alpha\oplus$d)	[43]
^{3}He(n,p)^{3}H	^{3}He(^{2}H,pp)^{3}H	^{2}H = (p$\oplus$n)	[21]
^{7}Be(n,α)^{4}He	^{7}Be(^{2}H,$\alpha\alpha$)n	^{2}H = (p$\oplus$n)	[44, 45, 46]

Various measurements, with a particular focus on lower energies, were carried out by different research groups. Batchelor et al. [48] conducted direct measurements within the energy range of 0.1$\leq E_{cm} \leq$1 MeV, Gibbons et al. [49] performed inverse measurements, and Costello et al. [50] directly measured reactions in the range of 0.3$\leq E_{cm} \leq$1.1 MeV. Additionally, theoretical predictions, including the most recent work by Drosg et al. [51], covered a wider energy spectrum. For astrophysical applications, reaction rates were subsequently calculated by Brune et al. [52], Adahchour et al. [53], and Smith et al. [54], all of which exhibit a consistent trend at astrophysically relevant temperatures. In contrast, Caughlan & Fowler [55] calculated a significantly higher reaction rate.

In the energy range under consideration, the available data are quite limited and predominantly date back more than half a century, primarily due to formidable experimental challenges. Consequently, these measurements have often yielded errors of up to 30%, depending on the energy level.

The experiment made use of a ^{3}He beam with a total kinetic energy of 9 MeV, generated by the FN Tandem accelerator at the Nuclear Physics Laboratory of the University of Notre Dame. Detailed information about the experimental setup can be found in [21].

Following the standard THM procedures, the experimental data were compared to previously published data from the literature after appropriate normalization (within the energy range of 350 to 400 keV). The resulting outcomes are presented in Table 3. These results demonstrate a strong concurrence with both direct and indirect measurements already documented in the scientific literature, reaffirming the validity of the THM approach for neutron-induced reactions.

3. Discussion and Perspectives

The reaction rates for four critical reactions within the Big Bang Nucleosynthesis (BBN) network, spanning the temperature range (0.001 $\leq$ T$_9$ $\leq$ 10), namely, ^{2}H(d,p)^{3}H, d(d,n)^{3}He, ^{3}He(d,p)^{4}He, and ^{7}Li(p,α)^{4}He, have been computed numerically, taking into account recent

Table 3. Cross section for the process of interest normalized to direct data as discussed in the text, together with the center-of-mass energy as reported in [21]. The related error is also reported.

E_{cm} (MeV)	σ (barn)	error on σ(b)
0.005	10.4	1.0
0.035	3.8	0.3
0.065	2.4	0.2
0.095	2.0	0.2
0.125	1.8	0.2
0.155	1.5	0.2
0.185	1.2	0.1
0.215	1.2	0.1
0.245	1.0	0.1
0.275	1.0	0.1
0.305	1.0	0.1
0.335	0.8	0.1
0.365	0.8	0.1

THM measurements. The uncertainties in experimental data, both from direct measurements and THM data, have been thoroughly incorporated for these reactions. The application of a similar approach to other reactions within the BBN reaction network will be explored in an upcoming publication.

The parameters governing each reaction rate are documented in [15]. The derived reaction rates are compared with some of the commonly employed compilations found in the literature, which are used to compute the abundances of ^{3,4}He, deuterium (D), and ^{7}Li in the context of BBN. The resulting abundances are found to be consistent, within the experimental uncertainties, with those obtained using the compilation of directly-measured reaction rates.

These findings underscore the efficacy of the THM as a valuable tool for investigating charged particle-induced reactions at energy levels characteristic of BBN. Additional efforts in examining reactions amenable to investigation via THM, including ^{3}He(n,p)^{3}H and ^{7}Be(n,p)^{7}Li, have been initiated, further solidifying the role of THM in the field of primordial nucleosynthesis. A new dataset, which will extend the explored energy range up to 800 keV, is currently undergoing analysis. This extension will enhance the accuracy of the related reaction rates and shed light on potential astrophysical implications.

Nonetheless, these indirect method measurements, in conjunction with numerous others, particularly concerning the ^{7}Be+n interaction [56] and references therein, suggest the possibility of a non-nuclear solution to the cosmological lithium problem. In this context, additional investigations into the destruction of lithium in stars, with particular attention to population II stars, may be warranted.

Acknowledgements
We thank the staff of the LNS target laboratory, for target preparation as well as the Nuclear Science Laboratory (supported by grant the National Science Foundation under grant number NSF PHY-1713857) staff at the University of Notre Dame for their invaluable efforts.

References

[1] G. Steigman 2007, Ann. Rev. Nucl. Part. Sci. 57, 463
[2] B. D. Fields and S. Sarkar 2006, J. Phys. G33, 220
[3] G. Israelian 2012, Nature 489, 37
[4] E. Komatsu et al. 2011, Ap. J. Sup. 192, 18
[5] E.W. Kolb and M.S. Turner 1990, "The Early Universe", Addison-Wesley
[6] D. Ezer and A.G.W. Cameron 1963, Icarus 1, 422
[7] C. Iliadis 2007, "Nuclear Physics of Stars", Wiley
[8] C. Casella et al. 2002, Nucl. Phys. A706, 203
[9] G.Baur, C.A. Bertulani & H Rebel , Nucl. Phys. A **458**, (1986) 188
[10] G.G. Kiss, M. La Cognata, C. Spitaleri et al., (2020) Phys. Lett. B, 135606
[11] C.Spitaleri 1991, in *Problems of Fundamental Modern Physics, II* : Proceedings, Ed. by R. Cherubini,P. Dalpiaz, and B. Minetti (World Sci., 1991), p. 21
[12] Spitaleri, C.; La Cognata, M.; Lamia, L.; Pizzone, R. G.; Tumino, A.; *Eur. Phys. Journ. A* **2019**, *55*, 161.
[13] A. Rinollo et al., Nucl. Phys. A **758**, (2005)146c
[14] A. Tumino et al., Phys. Lett. B **705**, (2011) 545
[15] R.G. Pizzone et al., Astrophys. J. **786**, (2014) 112
[16] Aliotta, M., Spitaleri, C., Lattuada, M., et al. 2000, European Physical Journal A, **9**, 435
[17] R.G. Pizzone et al., A. & A. **438**, (2005) 779
[18] L. Lamia et al., Astrophys. J. **768**, (2013) 65
[19] A. Tumino, et al., Ann. Rev. Nucl. Part. Sc. **71**, 345 (2021)
[20] A. Tumino et al., in: Handbook of Nuclear Physics (Springer, Singapore, 2022)
[21] R.G. Pizzone, C. Spampinato et al., Eur. Phys. J. , **A56**, 199 (2020)
[22] S. Romano, et al., Eur. Phys. Jour. A **27**, (2006) 221
[23] L. Lamia et al., Astrophys. J. **811**, (2015) 99
[24] Indelicato, I., La Cognata, M., Spitaleri, C., et al. 2017, The Astrophysical Journal, **845**, 19
[25] L.Lamia, et al., J. Phys. G 39, (2012) 015106
[26] Pizzone, R. G., D'Agata, G., La Cognata, M., et al. 2017, The Astrophysical Journal, **836**, 57
[27] D'Agata, G.; Pizzone, R. G. et al. 2018, The Astrophysical Journal, **860**, 61
[28] Sergi, M. L., Spitaleri, C., La Cognata, M., et al. 2015, Phys. Rev. C, **91**, 065803
[29] Pizzone, R. G., Roeder, B. T., McCleskey, M., et al. 2016, European Physical Journal A, **52**, 24
[30] La Cognata, M., Pizzone, R. G., José, J., et al. 2017, The Astrophysical Journal, **846**, 65
[31] Guardo, G. L., Spitaleri, C., Lamia, L., et al. 2017, Phys. Rev. C, **95**, 025807
[32] R.G. Pizzone et al. 2013, Phys. Rev. C **87**, 025805
[33] R.G. Pizzone et al., Phys. Rev. C **80**, (2009) 025807
[34] L.Lamia,M. La Cognata, C. Spitaleri, B. Irgaziev, R.G. Pizzone, Phys. Rev. C **85**, (2012) 025805
[35] Pizzone R.G. et al., Eur. Phys. J. A, **56** 283 (2020)
[36] R.G. Pizzone et al. 2005, A. & A. 438, 779
[37] L. Lamia, et al. 2012, Phys. Rev. C 85, 025805
[38] L. Lamia et al. 2017, Ap. J. 850, 175
[39] M.S. Smith, L.H. Kawano and R.A. Malaney 1993, Ap. J. 85, 219
[40] R.H. Cyburt 2004, Phys. Rev. D 70, 023505
[41] A. Coc, S. Goriely, Y. Xu, M. Saimpert, and E. Vangioni 2012, Astrophys. J. 744, 18
[42] L. Lamia et al. 2012, A. & A. 541, 158
[43] La Cognata, et al., 2005, Phys. Rev. C, **72**, 065802
[44] L. Lamia et al., Ap. J. **850**, 175, (2017)
[45] L. Lamia, M. Mazzocco, R.G. Pizzone, et al. , Astrophys. J., **879**, 23 (2019)
[46] S.Hayakawa, M. La Cognata, L. Lamia et al., Astroph. J,**915**, L13, (2021)
[47] J.H. Coon 1950, Physical Review 80, 488.
[48] R. Batchelor, R. Aves, and T.H.R. Skyrme 1955, Review of Scientific lnstruments 26, 1037.
[49] J.H. Gibbons and R.L. Macklin 1959, Physical Review 114, 571.

[50] D.G. Costello, S.J. Friesenhahn, and W.M. Lopez 1970, Nuclear Science and Engineering 39, 409.
[51] M. Drosg and N. Otuka 2015, INDC (AUS)-0019.
[52] C.R. Brune, K.I. Hahn, R.W. Kavanagh, and P.R. Wrean, 1999, Physical Review C60, 015801.
[53] A. Adahchour and P. Descouvemont 2003, Journal of Physics G: Nuclear and Particle Physics 29, 395.
[54] M.S. Smith, L. H. Kawano, and R. A. Malaney 1993, The Astrophysical Journal Supplement Series 85, 219.
[55] G.R. Caughlan and W.A. Fowler 1988, Atomic Data and Nuclear Data Tables 40, 283
[56] L. A. Damon et al., 2018, Phys. Rev. Lett., 121, 042701

Clustering states in neutron-rich nuclei

F. Risitano[1,2], B. Gnoffo[1,3], M. Trimarchi[1,2], L. Acosta[4], G. Cardella[1], E. De Filippo[1], D. Dell'Aquila[5,6], E. Geraci[1,3], I. Lombardo[1,3], C. Maiolino[7], N.S. Martorana[3,7], A. Pagano[1], E.V. Pagano[7], M. Papa[1], S. Pirrone[1], G. Politi[1,3], F. Rizzo[3,7], P. Russotto[7], G. Santagati[1] and C. Zagami[3,7,8].

[1] INFN - Sezione di Catania, Catania, Italy,
[2] Dipartimento MIFT, Università degli Studi di Messina, Messina, Italy,
[3] Dipartimento di Fisica e Astronomia "Ettore Majorana", Università degli Studi di Catania, Italy,
[4] Instituto de Física, Universidad Nacional Autónoma de México, Ciudad de México, Mexico,
[5] Dipartimento di Fisica "Ettore Pancini", Università degli Studi di Napoli Federico II, Italy,
[6] INFN - Sezione di Napoli, Napoli, Italy,
[7] INFN - LNS, Catania, Italy,
[8] Centro Siciliano di Fisica Nucleare e Struttura della Materia, Catania, Italy.

E-mail: fabio.risitano@ct.infn.it, fabio.risitano@unime.it

Abstract. The study of clustering states in neutron-rich nuclei is an important subject of research in the field of nuclear physics, steadily growing in interest in the international scientific community. In this context, break-up reactions play an important role for the characterization of exotic states in radioactive light nuclei, like neutron halos around stable cores, α-clustering structures or exotic clusters. The CLIR (Clusters in Light Ion Reactions) experiment was performed at INFN - Laboratori Nazionali del Sud (LNS), aiming at the investigation of such states in light radioactive nuclei, by producing a radioactive beam at the FRIBs facility. Reaction products were detected by the CHIMERA multidetector, coupled with four telescopes of the FARCOS array. Calibrations of the tagging system and of the FARCOS telescopes have been performed, for which accurate procedures have been carried out. In this paper, results on the analysis will be presented. Moreover, a brief review on the new fragment separator FRAISE, currently under construction at LNS, will be given.

1. Introduction

Clustering is an important phenomenon in the context of light neutron-rich nuclei as it represents a very powerful tool to investigate the behaviour of nuclear forces. It is in fact well known that in α-conjugated nuclei (like ^{8}Be, ^{12}C, ^{16}O, ^{20}Ne), nucleons can group together into cluster structures of α particles [1, 2]. This behaviour leads to resonance effects with even highly deformed cluster structures, rather than compact configurations, which have been predicted [3] at several energies and can have remarkable involvements in different fields, like astrophysics [4] or the study of the Equation of State (EOS) of nuclear matter. The ^{8}Be case is quite emblematic as it presents a resonance state at about 92 keV, making the ground state unbound [5, 6], showing a deformed dumbbell shape of α particles; moreover, it also plays important roles in the production of ^{12}C in the stars [7]. Similarly, the Hoyle state in ^{12}C is thought to be a weakly-bound state of α particles in a triangular shape, playing an important role in the carbon

Journal of Physics: Conference Series **2619** (2023) 012013 doi:10.1088/1742-6596/2619/1/012013

synthesis in the final stage of evolution of the stars [7, 8, 9, 10]. In recent years, many studies have also investigated the structures of non-self conjugated nuclei, showing that the addition of neutrons provides a sort of covalent bond between α clusters, increasing their stability. These molecular structures, kept together by so called valence neutrons, in analogy to electrons in molecular bonds, lead to the high deformations and peculiar shapes, characterized also by the appearance of rotational bands. At INFN-LNS, since the early 2000s, it has been possible to produce radioactive ion beams (RIBs) by the In-Flight fragmentation technique by means of the FRIBs facility [11]. In this context, very interesting study cases have been investigated [12], like the ^{4}He+^{6}He break-up channel of ^{10}Be [13], by means of the CHIMERA experimental apparatus [14, 15, 16], in which a new possible state at 13.5 MeV (6^+) has been investigated. Furthermore, a new experiment CLIR (Cluster in Light Ion Reactions) [17] was done in order to study also the break-up of ^{13}B and ^{16}C, increasing the intensity of the primary beam and improving the experimental apparatus. Moreover, the line of research on cluster states in neutron-rich ions will continue in the forthcoming years, thanks in particular to the construction of the new fragment separator FRAISE [18, 19], producing RIBs with much higher intensities.

2. Experimental setup of the CLIR experiment

The CLIR experiment was performed at INFN-LNS [17], where a radioactive ion beam was produced at the FRIBs (in-Flight Radioactive Ions Beams) facility through the In-Flight method [11, 20]. A primary beam of $^{18}O^{8+}$ at 55 MeV/u, accelerated by the Superconducting Cyclotron, was fragmented on a 1.5 mm thick beryllium target. Isotopes resulting by the fragmentation were then selected by the fragment separator FRIBs with a rigidity $B\rho \approx 2.8$ Tm and a momentum acceptance of $\Delta p/p \approx 1\%$, optimized for the transport of ^{11}Be. The isotopes of the cocktail beam were identified by using the ΔE-ToF technique [21]: a microchannel plate (MCP) detector measured the Time-of-Flight (ToF) start while a Double Sided Silicon Strip Detector (DSSSD) 156 μm thick was used both as ToF stop and energy loss ΔE, placed $\sim$ 12.9 m apart. The experimental data was gathered by the CHIMERA 4π multidetector [14, 15], coupled with four FARCOS telescopes [16, 22, 23]. CHIMERA is made of 1192 telescopic units, each one made of a Silicon detector $\sim$ 300 μm thick followed by a CsI(Tl) crystal, 3 to 12 cm thick, depending on the angular position of the unit. Telescopes are arranged in 9 rings, covering the forward angles from 1° to 30°, while a sphere covers the polar angles $30° \leq \theta \leq 176°$, for a total of $\simeq 94\%$ of the whole solid angle. A CH_2 50 μm thick polyethylene target was placed at the centre of CHIMERA's sphere. Moreover, CHIMERA was coupled with four FARCOS telescopes, placed between the sphere and the rings, covering polar angles from 1.6° to 8.5°. FARCOS is made of three different stages, two DSSSD (32×32 strips front/back) 300 μm and 1500 μm thick respectively, followed by four CsI(Tl) scintillators 6 cm thick, readout by a photodiode. The FARCOS telescopes show excellent characteristics of angular and energetic resolutions thanks to the high granularity of the first two strip stages, fundamental for the study of particle-particle spectroscopy in which high precision in the reconstruction of break-up products is required.

3. Calibration procedures

Calibration of the detectors is essential to reach reliable results. Through the calibrated tagging system it is possible to compare experimental data with simulations, and determine the composition of the cocktail beam used during the experiment. Moreover, the energy loss on the DSSSD detectors of FARCOS is employed to determine the total kinetic energy of the break-up fragments of interest, and so to reconstruct the excitation energy, through the invariant mass method, of the original cluster nucleus.

The three stages of the FARCOS array were calibrated individually using different techniques, depending on the stage. The 1500 μm thick stage was calibrated by calculating the cocktail beam ions energy loss on the stage, undergoing elastic scattering on the carbon ions of the

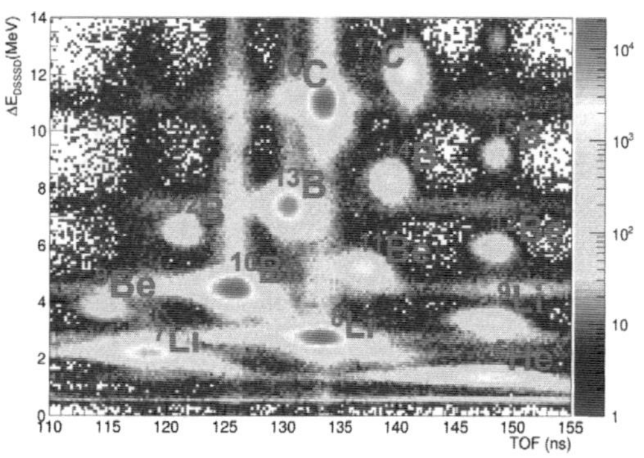

Figure 1. Tagging ΔE-ToF plot, calibrated via simulation. All the isotopic species of the cocktail beam have been identified [17].

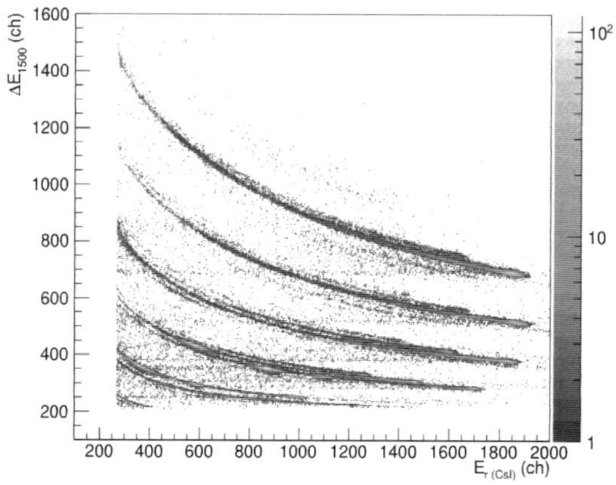

Figure 2. $\Delta E_{1500} - E_{CsI}$ plot obtained from the data of a single pixel for the 1500 μm stage and the corresponding CsI(Tl) scintillator. Good resolution of the ridges of each isotopic species is achieved.

reaction target. Then, correlating these values with the corresponding channel evaluated from $\Delta E_{1500} - E_{CsI}$ plots (Fig. 2), for each of the 32 different pixels of each strip, allowed to obtain a linear calibration as shown in the example in Fig. 3. Fig. 4 shows for each point the fit in Fig. 3 the relative error, that is generally less than 1%.

On the contrary, in the calibration of the CsI(Tl) stage, the light emission in the CsI scintillator can have a non-liner dependence on energy, charge and mass of the impinging ion. From the energy loss on the second 1500 μm thick stage, the value of the residual energy is calculated. Then, starting from the ΔE_{1500} calibrations, it is possible to associate the channel of the energy

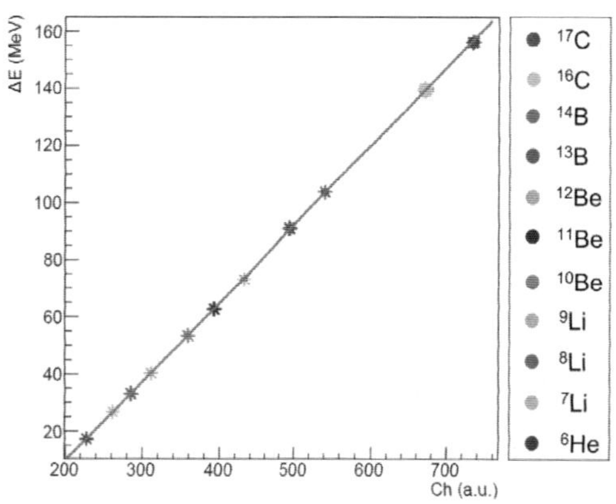

Figure 3. Linear calibration relative to a single strip of a 1500 μm stage of a FARCOS detector.

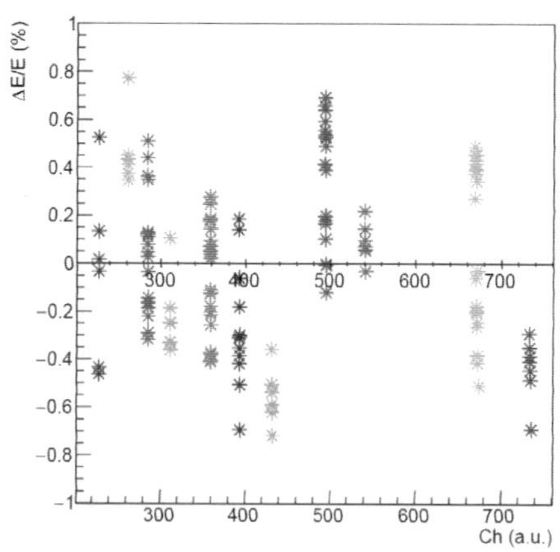

Figure 4. Relative error (%) for each point used to calibrate a single DSSSD strip, for the calibration shown in Fig. 3.

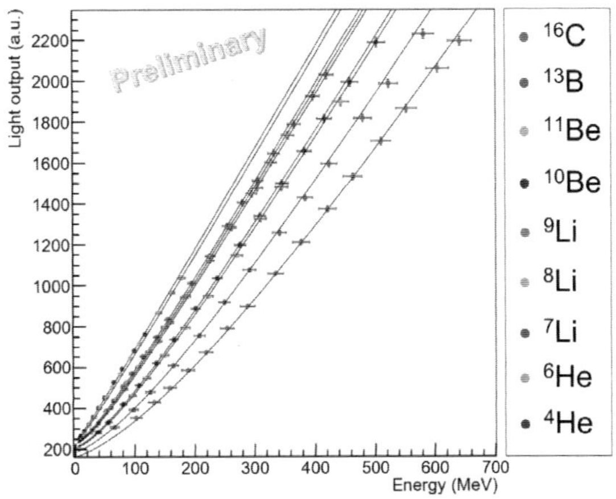

Figure 5. Calibration obtained for a CsI(Tl) crystal of the third stage of a FARCOS detector, using the RNQM model as fit function.

deposited on the CsI detector with the correspondent residual energy in MeV, evaluated from the second stage calibration (Fig. 5). Points are thereafter fitted using a multi-fit chi-square minimization technique, using the RNQM model [24], from equation 1, from which calibration parameters for each CsI can be estimated:

$$L = a_0 + a_1 E - a_1 a_2 A Z^2 \ln \left| \frac{E + a_2 A Z^2}{a_2 A Z^2} \right| + a_2 a_4 A Z^2 \ln \left| \frac{E + a_2 A Z^2}{a_3 A + a_2 A Z^2} \right|; \quad (1)$$

where L is the light output, A, Z and E are the mass, charge and energy of the incident ion, while parameters from a_0 to a_4 are fit parameters.

Calibration of the FARCOS 300 μm thick stage was done by calculating, event by event, the incident energy on the 1500 μm thick second stage from its calibrated energy loss value. In this way, the energy loss value on the first stage can be calculated, and so by correlating it with the correspondent value in channels, a correlation plot, as the one in Fig. 6 was produced for each strip of the first detector, from which a linear calibration can be carried out.

Fig. 7 shows the goodness of the overall calibration procedure, by overlapping the calibrated experimental data with a simulation of the energy losses for ^{17}C, ^{16}C, ^{13}B, ^{10}Be and ^{8}Li ions: a good comparison can be appreciated, demonstration of the goodness of the calibrations produced.

4. Preliminary result on ^{10}Be spectroscopy

In this section, preliminary results on the reconstruction of the excitation energy of ^{10}Be for the ^{4}He $+^{6}$ He break-up channel are shown. This is done by analysing the coincidence fragments coming from the de-excitation of ^{10}Be after being inelastically excited by interacting with the hydrogen of the target. Measuring the energy and emission angle of the ^{4}He and ^{6}He fragments and reconstructing their 4-momenta, it is possible, through the invariant mass technique [25], to obtain the kinetic energy of the sum of the two emitted ions in the ^{10}Be nucleus frame of reference. The excitation energy of ^{10}Be is then obtained by summing the energy threshold E_{th} = 7.409 MeV of the α emission. Fig. 8 shows the reconstructed invariant mass spectrum: from the plot one can see a good evidence of the 7.5 MeV and 9.5 MeV states [1] and a possible further confirmation of the 13.5 MeV 6^{+} state, already identified at LNS during previous experiments [13] by the CHIMERA collaboration. Further improvements are

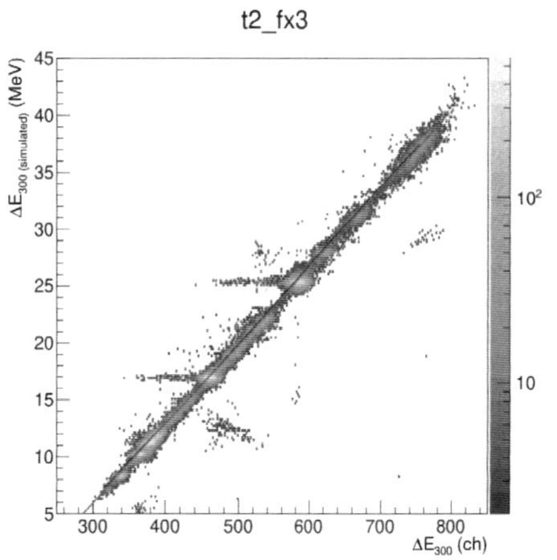

Figure 6. Calibration obtained for a single strip of the 300 μm stage of a the FARCOS detector, by simulating the energy loss from the 1500 μm stage.

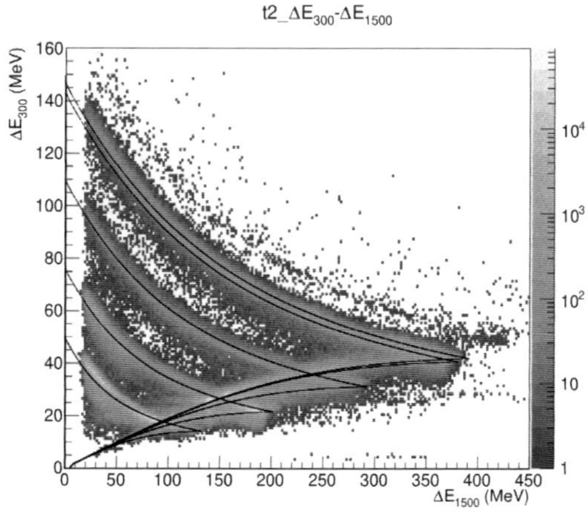

Figure 7. Calibrated $\Delta E_{300} - \Delta E_{1500}$ plot overlapped with simulations of energy loss for many ions of the cocktail beam.

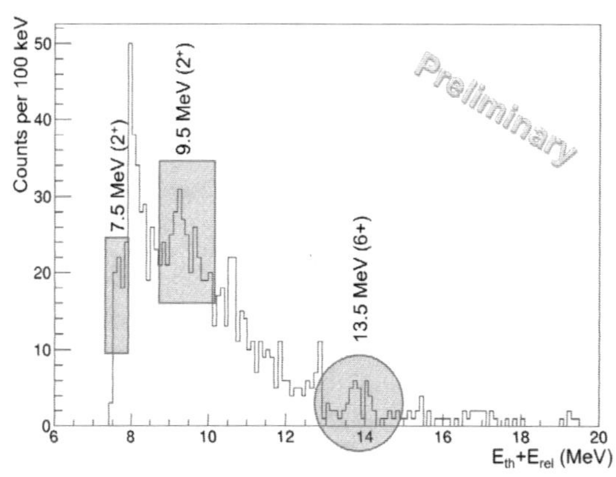

Figure 8. ^{10}Be relative energy spectrum obtained for the ^{4}He+^{6}He break-up channel. Location of known peaks is shown, at 7.5 MeV and 9.5 MeV, while possible evidence of the 13.5 MeV (6^+) state is shown.

needed, in particular by reducing background noise with event mixing procedures and removing the component of reaction with the ^{12}C of the polyethylene target, or for example deeply investigating events of dubious origin as the peak at 8 MeV. Statistics could be further increased by taking into account the recoil hydrogen of the reaction target, detected with the CHIMERA array.

5. The new fragment separator FRAISE

The study of cluster physics in neutron-rich nuclei at LNS will be further expanded in the next few years, thanks to the implementation of the new POTLNS project [11, 18, 19], currently under development. The project will provide an upgrade of the Superconducting Cyclotron

(CS), that will allow to produce beams of power up to 2-10 kW, for masses up to $A \approx 40$ in the range of Fermi energies. Moreover, the new fragment separator FRAISE will be built, that will exploit primary beam intensities up to 3 kW for the production of RIBs in the order of $10^3 - 10^7$ pps, resulting in a increase in yield of a factor of ≈ 20 compared to the beams produced by the previous FRIBs facility. FRAISE will be made of 4 dipoles (maximum $B\rho \approx 3.2$ Tm), 6 quadrupoles and 6 sextupoles, arranged as in Fig. 9 (right) in two symmetric branches, to fulfil optimal achromatic conditions of transport of the beam and reduce aberration effects. A new CLIM fragmentation target [26] will be installed, consisting of beryllium or carbon rotating discs of different thickness, equipped with a water cooling system and a robotic system to pick up and store different targets. Two slits will be installed: the first, located at the centre of the spectrometer, corresponding to the point with the highest energy dispersion, will be used to vary the momentum acceptance $\Delta p/p$, required also for the production of energy focused beams for future NUMEN experiments [27]; the second one, located at the exit of the spectrometer, will be used to reduce unwanted components or impurities of the cocktail beam. Moreover, aluminium degraders of different thicknesses can be also installed at the centre of the spectrometer, after the central slit, in order to produce purer secondary beams. Due to the new higher intensities of the primary and radioactive beams, it will be necessary to develop a new tagging and diagnostics system for FRAISE. This will have to be able to sustain the high beam intensities for long periods of time and will have to be equipped with a fast and compact readout system. For this reason, a new detection system based on silicon carbide (SiC), chosen for its radiation hardness features, is under development [19, 28]. The new detectors will be made of two arrays, each made of 6×12 single pads, arranged as in Fig. 10, 5×5 mm^2 and 100 μm thick, for a total of 72 channels and a total active area of 60×30 mm^2. A new fast frontend electronics is also currently under development, aiming to obtain a high timing resolution of the order of 100 ps. These double-array detectors will be placed: the first at the centre of the spectrometer, after the central slit and degrader; the second at the exit chamber, after the exit slits.

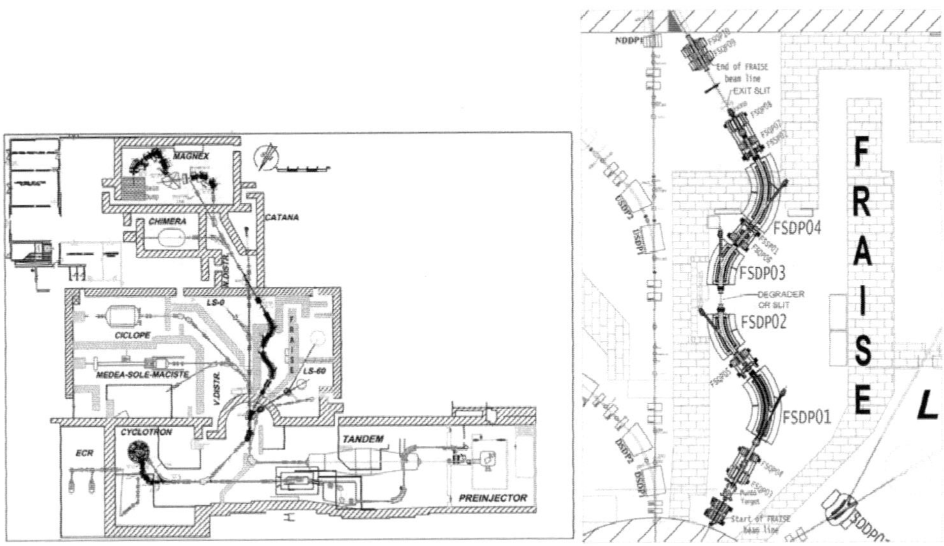

Figure 9. Map of the INFN-LNS facility (left), in which the position of the FRAISE apparatus is shown, with the relative scheme (right) [18]. Lables FSDP, FSQP and FSSP indicate the position of dipoles, quadrupoles and sextupoles, respectively.

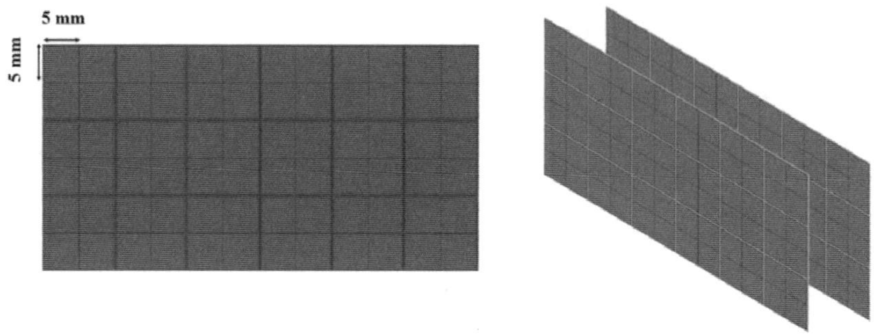

Figure 10. Scheme of the detectors for the new tagging and diagnostics system in development for the FRAISE separator, based on SiC multi-pads [19].

5.1. New possibilities with FRAISE

The new fragment separator FRAISE will give the possibility to produce, other than radioactive beams with higher intensities in the Fermi energy regime, new RIBs of acceptable intensity with more extreme isospin values, which could not be produced with the previous FRIBs facility. Many research topics already explored in the past years at LNS can be deepened, as well new ones can be studied. Systematic LISE++ simulations have been conducted [19, 29, 30], showing the new capabilities of the new spectrometer, both to simulate the theoretical production yield and many characteristics of the beam like horizontal and vertical apertures, spatial distribution of the various components and ΔE-ToF plots to identify the cocktail beam. Simulations were done considering the final FRAISE configuration, also including a 100 μm thick aluminium degrader, placed at the centre of the spectrometer on the symmetry plane, and a 100 μm thick SiC detector placed after the exit slit, used as tagging detector. Primary beams simulated had a power up to 2 kW, with an intensity lower than the maximum achievable by the new CS because of radiation protection issues. Moreover, the thickness of the ^{9}Be fragmentation target is optimized for the production of the ion of interest, while the rigidity and slits of FRAISE are optimized for the transport and for the production of a purer beam. With FRAISE it will be possible to further widen the topic of clustering structures, as it will be possible to increase the yield of the production up to a factor or 20 with respect to FRIBs, also having the possibility to produce purer beams, both for study cases already investigated at LNS (this is the case of ^{10}Be and ^{16}C [13, 17]), as well as new cases like ^{13}B [12]. Moreover, among other topics of research there are halo structures, both neutron-rich (^{11}Be) [31] or neutron-poor ^{8}B [32], for which interesting phenomena are theorized, or deepen the investigation of the Pygmy Dipole Resonance in a vast range of nuclei, like ^{20}O, ^{34}Si, ^{38}S, ^{48}Ar and ^{68}Ni. In this context the FRAISE facility will be a competitive facility in the international panorama of nuclear physics at the Fermi energy regime.

6. Conclusions

In this paper, a report on the analysis of cluster physics at LNS and in particular of the CLIR experiment was given. The CLIR experiment, performed at the FRIBs facility, aimed at the study of cluster structures in light neutron-rich nuclei, studied through the CHIMERA multidetector and FARCOS array. An overview of the calibrations of various detectors was given. The good performances of the tagging system were shown, able to identify the components of the cocktail beam and through which it is possible to conduct an analysis aimed specifically at a single ion of interest. Moreover, the FARCOS detectors were calibrated, showing good results in the resolutions and isotopic separation in $\Delta E_{1500} - E_{CsI}$ and $\Delta E_{300} - \Delta E_{1500}$ matrices. These data were used for a preliminary work on ^{10}Be spectroscopy, in particular for the ^{4}He+^{6}He break-up channel, which will be improved by increasing the statistics and reducing the background. Moreover, the research topic of cluster structures will be deepened in the next years of service at LNS with the construction of the new facility FRAISE for the production of radioactive beams. In fact, the new facility will provide the possibility to produce RIBs of greater intensity with respect to the previous FRIBs facility, allowing to widen the topics of research by including more unstable isotopic species, far from the stability valley. Simulations on LISE++ have proven the excellent possibilities of the new production system. Moreover, a campaign of tests and simulations on the new diagnostics and tagging system based on SiC technology has started. Finally, the experimental apparatus will also be improved, in particular by adding the new NArCOS hodoscope [33, 34, 35], currently under development by the CHIRONE collaboration, that will provide information on the neutron emissions in cluster break-up reactions allowing for the reconstruction of complete events.

References

[1] M. Freer, Rep. Prog. Phys. 70, 2149 (2007).
[2] L. R. Hafstad and E. Teller, Phys. Rev. 54, 681 (1938).
[3] K. Ikeda, N. Takigawa, and H. Horiuchi, Prog. Theor. Phys. Suppl. E68, 464 (1968).
[4] P. Descouvemont, Jour. Phys. G, 35 (2008) 014006.
[5] C. Cardella et al., Eur. Phys. J. Plus, 138 (2023) 89.
[6] D. Tilley et al., Nucl. Phys. A, 745 (2004) 155.
[7] F. Hoyle, Astrophys. J. Suppl. Ser. 1, 121 (1954).
[8] G. Cardella et al., Il Nuovo Cimento C, 45 (2022) 56.
[9] G. Cardella et al., Nucl. Phys. A, 1020 (2022) 122395.
[10] G. Cardella et al., Phys. Rev. C, 104 (2021) 064315.
[11] P. Russotto et al., Jour. of Phys.: Conf. Series, 1014 (2018) 012016.
[12] B. Gnoffo et al., Front. Phys. Sec. Nucl. Phys. 10, (2022), doi:10.3389/fphy.2022.1061633.
[13] D. Dell'Aquila et al., Phys. Rev. C, 93 (2016) 024611.
[14] A. Pagano et al., Nucl. Phys. A 734, (2004) 504.
[15] E. Geraci et al., Il Nuovo Cimento C, 45 (2022) 44.
[16] A. Badalà et al., La Rivista del Nuovo Cimento, 45 (2022) 189.
[17] F. Risitano et al. Il Nuovo Cimento C, 45 (2022) 60.
[18] A. Russo et al., Nucl. Instr. Meth. Phys. Res. Sec. B, 463 (2020) 418.
[19] N.S. Martorana et al., Front. Phys. Sec. Nucl. Phys. 10, (2022), doi:10.3389/fphy.2022.1058419.
[20] N.S. Martorana, et al., Il Nuovo Cimento C, 45 (2022) 63.
[21] I. Lombardo et al., Nucl. Phys. B, 215 (2011) 272.
[22] E.V. Pagano et al., EPJ Web of Conferences 117, (2016) 10008.
[23] L. Acosta et al., 2021 IEEE Nucl. Sci. Symp. Med. Img. Conf. (NSS/MIC), (2021) p. 1-3.
[24] M. Parlog et al., Nucl. Inst. Meth. A, 482 (2002) 693.
[25] A.G. Artyukh et al., Instrum. Exp. Tech., 52 (2009) 13.
[26] S. Grévy and R. Hue, 24th World Conference INTDS2008, Caen, France (2008).
[27] F. Cappuzzello, C. Agodi, D. Carbone, M. Cavallaro, Eur. Phys. J. A, 52 (2016) 167.
[28] C. Altana et al., 2021 IEEE Nucl. Sci. Symp. Med. Img. Conf. (NSS/MIC), (2021) p. 1–4.
[29] N.S. Martorana, Il Nuovo Cimento C, 44 (2021) 1.
[30] F. Risitano, Il Nuovo Cimento C, 45, (2022) 68.
[31] I. Talmi and I. Unna, Phys. Rev. Lett., 4 (1960) 469.
[32] T. Minamisono et al., Phys. Rev. Lett., 69 (1992) 2058.
[33] E.V. Pagano et al., Front. Phys. Sec. Nucl. Phys., 10, (2023).
[34] E.V. Pagano et al., Nucl. Instr. Meth. Phys. Sec. A, 889 (2018) 83.
[35] E.V. Pagano et al., Il Nuovo Cimento C, 45 (2022) 64.

Journal of Physics: Conference Series **2619** (2023) 012014 doi:10.1088/1742-6596/2619/1/012014

Total Reflection Xray Fluorescence (TXRF) spectrometry as a powerful and broad-spectrum analytical tool in the nuclear sciences

J L Rodríguez Alejandre[1], **Vijay R. Sharma**[2], **L. A. Acosta Sánchez**[3], **N Dasgupta-Schubert**[1,*]

[1]Facultad de Ciencias Físico Matemáticas, Universidad Michoacana de San Nicolás de Hidalgo, Morelia, México.
[2]Department of Radiation Oncology, School of Medicine, University of Maryland, Baltimore, MD, USA.
[3]Instituto de Física, Universidad Nacional Autónoma de México, Apartado Postal 20-364, 01000 México City, México

E-mail: *nabanita.schubert@umich.mx

Abstract. X-ray spectroscopy is widely used in nuclear reaction and structure studies. To aid such studies the technological envelope is continually being pushed to achieve higher quality such as superior power and brilliance, higher resolution and sensitivity. We present the discussion of a relatively new form of X-ray fluorescence spectrometry - Total Reflection X-ray Fluorescence (TXRF) spectrometry that can achieve analytical sensitivities of the order of parts per billion (ppb). An overview of the TXRF technique is presented and its utility for nuclear materials analysis discussed. Polarized Energy Dispersive X-ray Fluorescence (PEDXRF) spectrometry is also an XRF technique that can complement TXRFS where it may not be applicable. That scenario is discussed with respect to our results using PEDXRF on geothermal samples that serve to corroborate the recent findings of the Borexino Geoneutrino Experiment regarding the high Th, U concentrations in the earth's mantle.

1. Introduction

X-ray spectroscopy is widely used in nuclear reaction and structure studies, as for example x-ray isotope shifts for nuclear charge distributions [1, 2], the study of EC decay [3], mu-mesic atoms [4] and synchrotron processes [5], X-gamma-alpha coincidence spectroscopy for product Z identification in Super Heavy Nuclear synthesis [6] and x-ray bursts from thermonuclear processes in accreting neutron stars [7]. However, the most extensive area of the use of x-ray spectrometry in the nuclear sciences is in the quantitative analysis of trace elements in materials of importance to the nuclear industry and related areas. To aid such studies the technological envelope is continually being pushed to achieve higher quality such as superior power and brilliance of the x-ray source and higher resolution and sensitivity of the detection system. This work presents the discussion of a relatively new form x-ray fluorescence spectrometry (XRFS) that uses the property of the total reflection of a grazing x-ray beam on a perfectly reflective sample-containing substrate to achieve analytical sensitivities of the order of a nanograms in the case of laboratory TXRF (L-TXRF) systems and femtograms in the case of synchrotron radiation based TXRF (SR-TXRF) systems. Whilst undoubtedly SR-TXRF presents great analytical advantages, the compactness, portability, logistical and budgetary advantages of the L-TXRF as well as its allowance of in situ analysis have led to an accelerated growth of its usage. Much of this has been occasioned by rapid developments in L-TXRF instrumentation viz. the x-ray source the x-ray tube, better x-ray optics and detection systems. This work discusses the fundamentals of L-TXRF and some applications for nuclear materials analysis as well as the application of a related high sensitivity XRF technique, the Polarized Energy Dispersive XRF (PEDXRF) for the analysis of environmental radioactivity.

Ideally, the elemental analysis technique must be (a) accurate, (b) reproducible, (c) have a large dynamic range starting at the micro-trace level i.e. an analyte concentration of ≤ 1 ppm (1 mg/l), or an even lower level (ultra-trace), (d) have multi-element analysis capability, (e) require minimal sample mass,(f) sample preparation must be simple requiring minimal pre-processing, (g) be preferably non- destructive, (h) be low in initial and running costs, (i) be easy to maintain and (j) have a small footprint with an option of portability. It is clear that no one technique can meet all the aforesaid criteria. One purpose of this report is to examine the technique of TXRF Spectrometry (TXRFS) for the aforesaid criteria via a specific application (nuclear materials analysis). Several excellent reviews [8, 9, 10, 11] have appeared on TXRFS since the early 2000s. Most have focused on non-nuclear applications in. Indeed, TXRFS is nowhere as commonly known in the nuclear community as for instance, gamma ray spectrometry.

2. Fundamentals of TXRFS

Traditional XRFS suffers from significant Bremsstrahlung background and matrix absorption and scattering effects that produce relatively poor elemental sensitivity, making it unsuitable for trace level analysis. Starting with the late 1980s, a number of improvements of the fundamental XRF technique were achieved, e.g. TXRF, PEDXRF (polarized energy dispersive XRF), SRXRS (synchrotron radiation based x-ray spectrometries) including impressive advances in x-ray instrumentation. These enabled extending the sensitivity to the ultra-microtrace level (< 1 ppb or lower). TXRFS falls into the category of a large dynamic range, high-sensitivity micro-analytical technique for elemental analysis at the micro-trace level. The total reflection of x-rays (TRX) was discovered as early as 1923 by Arthur Holly Compton [12] but its exploitation for spectral analysis did not occur till the year 1971 when Yoneda and Horiuchi [13] used the TRX phenomenon on a flat reflector surface to simultaneously detect the masses of Cr, Fe, Ni and Zn at the order of 100 ng. The first commercial TXRF spectrometer was introduced by Seifert & Co., Ahrensburg, Germany and by Wobrauschek in Vienna, Austria both around 1980.

The underlying concept of the TXRF spectrometer is laid out in figure 1. The sample is placed as a thin homogeneous layer on the sample carrier. The incident x-ray beam enters at an angle below the critical angle and gets totally reflected at the interface of the sample+substrate and air. The fluoresced x- rays from the sample are collected by the detector placed above and very close to the sample, nearly orthogonal to the exciting beam.

Figure 2 shows the reflection and refraction of an x-ray beam as it hits a solid medium traversing through air. By Snell's Law,

$$n_1 \cos(\alpha_1) = n_2 \cos(\alpha_2) \tag{1}$$

where α_1 and α_2 are the angles of incidence and refraction with respect to the horizontal at the interface. For x-rays, all media are optically less dense than vacuum and all solids are optically less dense than air, so that $n_{solid} < n_{air} \approx 1$. In the TXRF spectrometer, the sample is deposited as a thin layer on a polished disc (usually of quartz) and the ambient is (normally) air. Thus $n_2 < n_1$ which implies that α_2 is always less than α_1, such that for a certain small value of α_1 called the critical angle, α_c, α_2 will become zero and the refracted x-ray will be

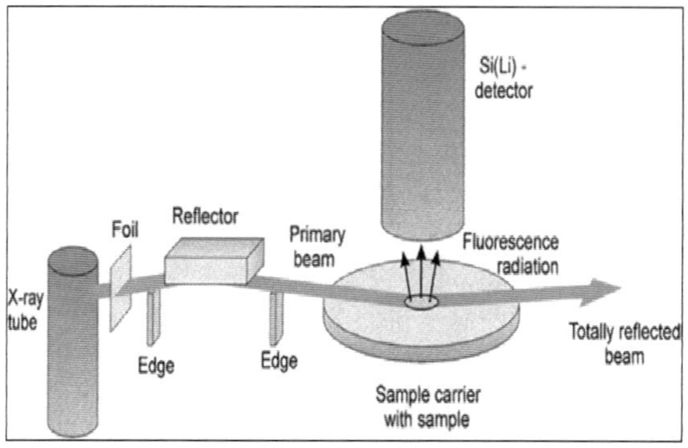

Figure 1. Concept and layout of a Total Reflection X-ray Fluorescence (TXRF) spectrometer. (From ref. [8]; reproduced with permission).

co-linear with the boundary of the interface. Substituting $n_1 = n_{air} \approx 1$ and $\alpha_2 = 0$ in equation 1, we get,

$$\cos(\alpha_c) = n_{solid} \tag{2}$$

n_{solid} is expressed as,

$$n_{solid} = 1 - \delta - i\beta \tag{3}$$

where $i = \sqrt{-1}$

The real component δ, called the decrement, is a measure of the dispersion. For x-rays it is of the order of 10^{-6}. It is expressed as,

$$\delta = \frac{N_A}{2\pi} r_o \lambda \frac{Z}{A} \rho \tag{4}$$

where N_A is the Avogadro number, r_o the classical electronic radius, λ the wavelength of the primary x - radiation, Z and A the atomic and mass numbers respectively of the element in the matrix and ρ the density of the matrix.

The imaginary component β is a measure of the absorption of the primary radiation and is given as,

$$\beta = \frac{1}{4\pi} \mu \lambda \tag{5}$$

where μ is the mass absorption coefficient. For most media it is even smaller than 10^{-6} and for transparent media $\frac{\beta}{\delta} \approx 0$. Utilizing equations 2 and 3 for transparent media we therefore obtain the critical angle as,

$$\alpha_c \approx \sqrt{2\delta} \tag{6}$$

Values of α_c are very low - usually small fractions of a rad. For example, for Mo-k_α x-rays passing from vacuum to quartz glass, $\delta = 1.5 \times 10^{-6}$, $\beta = 4.6 \times 10^{-9}$ and hence $\alpha_c = 0.0173$ rad or 5.9 min of arc. When the condition $0 < \alpha_1 < \alpha_c$ is implemented, Snell's law no longer yields real values for the angle of refraction. The interface behaves as a nearly ideal mirror fully reflecting the incident beam back into the first medium. The reflected beam can now interfere constructively with the incident beam, generating a region of x-ray standing waves (XSW) as shown in figure 3. This region occupies a space just above the surface of the reflector,

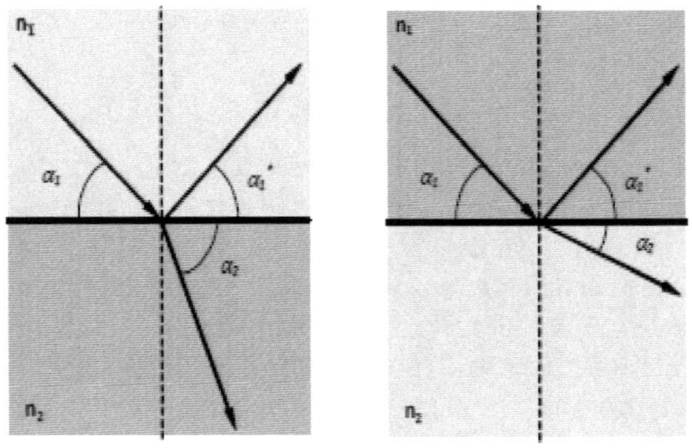

Figure 2. The reflection and refraction of the x-ray beam for two distinct cases of the refractive indices of the media, n_1 and n_2. (a) $n_2 > n_1$; (b) $n_2 < n_1$. α_1' is the reflected angle. (From ref. [14]; reproduced with permission).

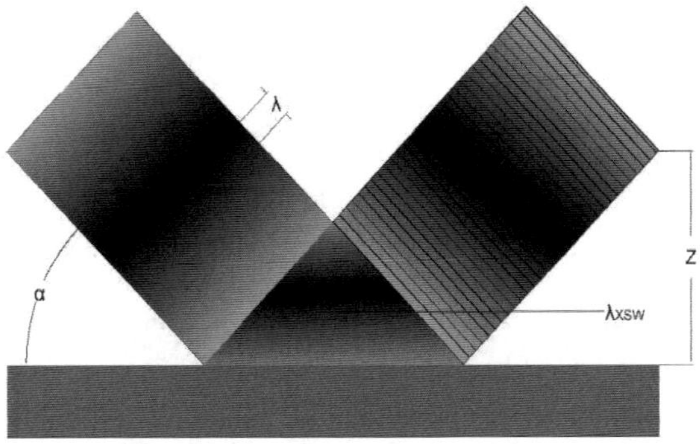

Figure 3. Total reflection at the surface of the reflector in TXRFS and the field of the XSW (see text). The triangle in the centre is the "magic triangle" that contains the idealised XSW. (From ref. [14]; reproduced with permission).

whose cross-section is termed the "magic triangle". For a thin sample deposited within the magic triangle, the presence of the XSW implies that the intensity of the exciting radiation is amplified. The atoms of the sample may be considered to be doubly excited - once by the incident x-ray beam and a second time by the totally reflected beam. Thus, on average, the TXRF geometry affords approximately double sample excitation as against that by a standard XRF set-up. Unlike XRF, in TXRF the spectral background is considerably reduced. Since the incoming primary x-ray beam gets totally reflected from the smooth reflector surface, there is very little contribution to the spectrum from the scattered and fluoresced radiation emanating from the material of the reflector.

Scattering and fluorescence from the sample matrix is also low because of the small volume of the sample material deposited on top of the reflector. In sum, the spectral background is reduced by 1:500 and excitation to fluorescence enhanced by 2:1 when comparing TXRFS to traditional XRFS. Consequently, analyte sensitivities are greatly enhanced leading to an absolute detection limit for transition metals of some picograms in L-TXRF units and some femtograms for SR-TXRF.

Following the description by Klockenkämper and von Bohlen [9], we express the intensity, $I_{fluo}(\alpha_0)$, of the fluorescent beam as seen by the detector, from an infinitesimally thin sample layer on an infinitely thick substrate as,

$$I_{fluo}(\alpha_0) \cong I_n C_A S_{x,E_o} \varepsilon_{det} T_{air} \cdot \left|E_1^t + E_1^r\right|^2 \tag{7}$$

Where I_n is a norm or reference for the fluorescence intensity registered by the detector, which is proportional to the primary beam intensity, I_0 ; C_A is the mass of the analyte element "x" per unit area of the sample layer; S_x,E_o is a sensitivity value of the element "x" at the photon energy E_o of the monochromatic primary beam ; ε_{det} is the efficiency of the detector for the fluorescent x-rays of energy E ; T_{air} is the transmission factor of the fluorescent x-ray photons through the air column between the sample and the detector ; E_1^t is the amplitude of the electric field vector of the beam refracted (or transmitted into the layer) at the interface between the thin sample layer and vacuum/air ; E_1^r is the amplitude of the electric field vector of the beam that is reflected (into the sample layer) at the interface between the thin sample layer and the thick substrate ; α_0 is the glancing angle of incidence of the primary beam which is assumed to be small ($< 5°$). A further simplification arises if the substrate is vacuum/air. In that case, the amplitude E_1^r nearly vanishes because the reflection at the bottom of the layer can be neglected. Equation 7 then reduces to the first term of the sum.

Figure 4 shows the variation of the fluorescent x-ray intensity of the analyte in a 500 nm thick sample deposited on a Si reflector, with respect to the angle of incidence α of the primary x-ray beam. Also shown is the intensity of the Si fluorescence radiation from the reflector which constitutes the background. The vertical red line at the position of the critical angle α_c demarcates the TXRF region at $\alpha < \alpha_c$ and the XRF region at $\alpha > \alpha_c$. In the TXRF region two effects are clearly noticeable. Firstly, the analyte signal is about double its value in the XRF region which, as stated earlier, is a consequence of the double excitation of the atoms of the analyte by the XSW field. Secondly, the background fluorescence intensity from the Si reflector drops down to very low values, at angles significantly lower than α_c. TXRF is a microanalytical technique since very small quantities of sample, typically 0.1 - 10 μg, are analysed. The sample is presented as a dry, homogeneous and evenly distributed deposit on the surface of the sample carrier (reflector), with a thickness between 0.1 - 10 μm. One of the principal advantages of TXRF as compared to conventional XRF or the ICP techniques is the simple quantification method which, moreover, can be achieved simultaneously for multiple elements. For an appropriately prepared sample deposited as a thin homogeneously distributed layer on the reflector, the calibration is linear. First a suitable internal standard whose concentration is known, is added to the sample and an aliquot of this composite sample is deposited onto the reflector.

The quantification uses the relative sensitivities for the element under study ("x") and for the internal standard ("int"), S_x and S_{int} respectively, the characteristic x-ray peak intensities of the element, I_x, and internal standard, I_{int}, the masses of the sample, m_{sample}, and internal standard, m_{int}, or the concentration (mass fraction) of the internal standard, c_{int}, in the linear equation,

$$c_x = \frac{I_x/S_x}{I_{int}/S_{int}} \frac{m_{int}}{m_{sample}} = \frac{I_x/S_x}{I_{int}/S_{int}} c_{int} \tag{8}$$

to determine the concentration of the element, c_x. With this simple quantification, the TXRF method is able to achieve a dynamic range in concentrations that spans 5 orders of magnitude from a few ppb (10^{-9}) to hundreds of ppm (10^{-4}).

The definition of the lower limit of detection (LLD) is,

$$LLD = \frac{3\sqrt{N_B}}{N_x} m_{sample} = \frac{3\sqrt{I_B}}{S_x} \frac{1}{\sqrt{t}} \tag{9}$$

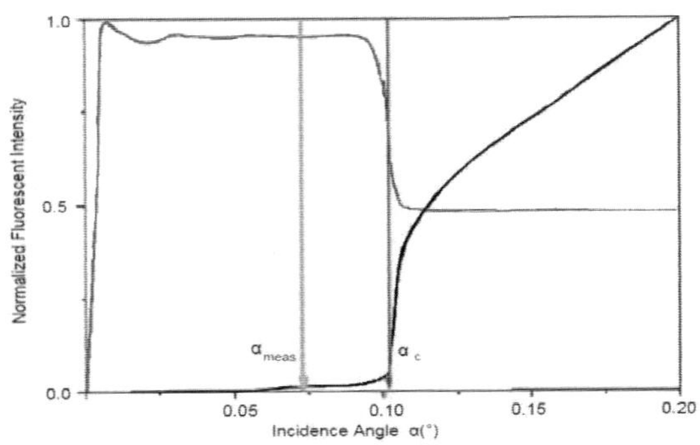

Figure 4. The variation of the analyte's fluorescence intensity (blue line) and the background intensity (black line) from the Si reflector, with respect to the angle of incidence α of the primary x-ray beam; α_c is the critical angle and α_{meas}, the actual angle of incidence. In a typical measurement, α_{meas} is set at 70% of the value of α_c. (From ref. [14]; reproduced with permission).

where N_x and N_B are the net counts of the peak and its background, m_{sample} the mass of the sample, I_B the intensity of the background, S_x the sensitivity of element "x" ($S_x = I_x/m_{sample}$ where I_x = intensity of "x"). SR- TXRF achieves ~3 orders of magnitude reduction in the LLD as compared to L-TXRF by substantially increasing I_x and reducing I_B. One disadvantage with XRF in general is the low inherent sensitivity for elements with Z $\leq$ 13 because of their low fluorescence yields and because their low-energy characteristic x-rays are attenuated by the intervening layers between the sample and the active volume of the detector. The inherent properties of the synchrotron reduce this limitation in SR-TXRF. As in all analytical techniques, certain artefacts may arise which give rise to errors in the estimation. In TXRFS these errors arise from three sources: instrumental, data processing and sample presentation. New instrumental and computational developments have improved the first two but the last remains an obstacle. Mori and Uemara [15] state that the sample depth distribution factor is an important and oft overlooked source of error.

3. Application of TXRFS and PEDXRFS to nuclear materials and environmental radioactivity analysis

Materials used in the nuclear industry constitute the precious, hazardous and radioactive, such as the nuclear fuel elements., Th, U, Pu, reactor poisons such as B and Cd, and the fuel rod cladding material "Zircalloy". Special safeguarding and handling processes are necessary in the analysis technique to (a) minimize material loss and afford radiation protection to instruments and personnel, concomitant to (b) fast, accurate and sensitive multielement analysis. The following features of TXRFS make it particularly attractive towards meeting those ends. (1) Requirement of small sample mass (μg or μL). This assures only a few Bq of radioactive exposure as well as reduces active waste generation. (2) Non-consumptive nature that minimizes loss. (3) Simple and fast analysis with low spectral interferences. (4) Negligible matrix and no memory effects. Hence no need for matrix matched standards which are difficult to prepare for nuclear materials. (5) Simple quantification procedure and simultaneous analysis across the periodic table. The last three features address in particular, the requirement (b) above. The oxidation state of U, Pu, Th determines the properties of their compounds, notably their oxides, used in the nuclear industry. It has been shown [16] that TXRFS can be successfully coupled to X-ray Absorption Near Edge Spectrometry (XANES) to determine the speciation of U oxides. A review of these topics may be found in [17].

Two of the principal elements that constitute natural radioactivity, are the elements Th and U. The determination of repositories of these elements are important for the nuclear industry as the first stage of their eventual use as nuclear fuels. Possibly, a relatively rich source of these elements could be geothermal and volcanic sites where elements from the earth's mantle are brought up to earth's crust. Recent results on geoneutrinos from the Borexino collaboration [18] indicate that the earth's mantle is relatively quite rich in Th and U. Accordingly it becomes necessary to explore these niche sites to confirm this hypothesis and corroborate indirectly the Borexino results. Abuhani et al have conducted a series of PEDXRF measurements on sediments and soil of the geothermal fumaroles of the Los Azufres volcanic caldera of Mexico [19, 20, 21] where they have found concentrations of Th, U that moderately exceed average values on the earth's crust. PEDXRF is an XRF technique where the monochromatized primary x- ray beam is doubly polarized at successive 90° angles to eliminate the Bremmstrahlung background and thereby achieve substantial improvements in analytical sensitivity. In this particular case, TXRFS would be inappropriate to use because of its low sample mass requirements. The low concentration of Th, U in the samples for low sample masses would be below the LLD of L-TXRF. PEDXRF on the other hand requires sample masses $\leq$ 4g making the total number of U and Th atoms large enough to be detectable. The results obtained by Abuhani et al [19, 20, 21] indicate an average Th concentration of more than the double of the value for the earth's crust and for U, an average concentration that is slightly lower possibly because of the run-offs of the more soluble uranyl carbonate complexes by weathering processes. In effect this investigation corroborates the inference of a higher mantle concentration of Th and U by the Borexino Neutrino Laboratory.

4. Conclusion

TXRFS and its complementary technique, PEDXRFS, are the modern extensions of the traditional XRF technique that have pushed the frontiers of the figures of merit of elemental analysis. The various beneficial features of TXRFS and its utility in the context of applications for nuclear materials analysis, are discussed. A case that illustrates the limitation of TXRFS is described through the application of PEDXRF to the analysis of geothermal ecological samples for concentrations of cadmium, mercury, lead, thorium and uranium.

5. Acknowledgments

JLRA thanks conacyt for the financial support in the form of scholarship.

References

[1] F. Boehm, P.L. Lee, Atomic Data and Nuclear Data Tables 14 (1974) R605-611.
[2] E.C. Seltzer, K X-Ray Isotope Shifts, Phys. Rev. 188 (1969) R1916.
[3] A. Palffy, Z. Harman, C. Kozhuharov, C. Brandau, C. H. Keitel, W. Scheid, T. Stohlker, Phys. Lett. B. 661 (2008) R330-334.
[4] A. Adamczak,A. Antognini, N. Berger, N. et al., Eur. Phys. J. A 59 (2023) 15.
[5] S. Kishimoto, Y. Yoda, M. Seto, Y. Kobayashi, S. Kitao, R. Haruki, T. Kawauchi, K. Fukutani, and T. Okano. Phys. Rev. Lett. 85 (2000) 1831.
[6] D. Ackermann, Nuclear Physics A 944 (2015) R376.
[7] J. J. M. int Zand, M. E. B. Visser, et al., A&A, 606(2017) A130.
[8] R. E. van Grieken, A. Markowicz, (Eds). 2002. Handbook of X-ray Spectrometry, 2nd Edition. Marcel Dekker Inc. New York - Basel.
[9] R. Klockenkamper, A. von Bohlen,2015. Total-Reflection X-Ray Fluorescence Analysis and Related Methods. Second Edition. In Chemical Analysis. A series of monographs on Analytical Chemistry and its applications, Vol. 181 (Ed.: M.F. Vitha). John Wiley & Sons, Hoboken, NJ, USA.

[10] B. B. Beckhoff, B. Kanngießer, N. Langhoff, R. Wedell, H. Wolff, (Eds). 2006. Handbook of Practical X-ray Fluorescence Analysis. Springer-Verlag, Berlin-Heidelberg.
[11] A. von Bohlen, Spectrochim. Acta Part B, 64 (2009) R821-832.
[12] A.H. Compton, Phil. Mag., 1121 C 45 (1923) 1121.
[13] Y. Yoneda, T. Horiuchi. Rev. Sci. Instr., 42 (1971) 1069.
[14] R. Fernandez-Ruiz, Development in Analytical Chemistry 1(2014)1.
[15] Y. Mori, K. Uemura, X-Ray Spectrom., 28(1999), R421-426.
[16] K. Sanyal, A. Khooha, G. Das, M.K. Tiwari, N.L. Mishra, Anal. Chem. 89(2017) 871-876.
[17] S. Dhara, N.L. Mishra., Trends in Anal. Chem. 116(2019) 31-43.
[18] M. Agostini et al. Phys. Rev. D 101, (2009).
[19] W.A. Abuhani, N. Dasgupta-Schubert et al., J. Env. Radioactivity. 139(2015) R33-42.
[20] W.A. Abuhani, N. Dasgupta-Schubert et al., J. Radioanal. Nucl. Chem. 322(2019) R1365-1371.
[21] W.A. Abuhani, N. Dasgupta-Schubert et al., 2021. "In the footsteps of Mendeleev: Heavy elements in three geothermal micro-ecological zones of the Los Azufres (Mexico) caldera." In Proceedings of the Conference, Particle Physics at the Year of 150th Anniversary of the Mendeleev's Periodic Table of Chemical Elements, pp. R568-578. World-Scientific.

Journal of Physics: Conference Series **2619** (2023) 012015 doi:10.1088/1742-6596/2619/1/012015

Structure of ^{13}C in the Cluster Shell Model

A H Santana Valdés[1], R Bijker[2]

Instituto de Ciencias Nucleares, Universidad Nacional Autónoma de México, Apartado Postal 70-543, 04510 Cd. de México, México

E-mail: [1]adrian.santana@correo.nucleares.unam.mx, [2]bijker@nucleares.unam.mx

Abstract. We study the structure of 13C in the framework of the Cluster Shell Model. A comparison of the available experimental data with our model is made. Some predictions for level ordering and form factors are presented.

1. Introduction

Recent measurements of rotational excitations in ^{12}C have renewed interest in the cluster structure of light nuclei, particularly α conjugate nuclei as clusters of k α-particles. The interest in these structures dates as back as 1937, with the works of [1–4]. The intricate patterns in which protons and neutrons are arranged give rise to a multitude of special properties, many of them of great implications, from nuclear physics to astrophysics, where the best example of this is the Hoyle state. Clustering in nuclei is complex, and the underlying physics is not yet fully understood, being a major challenge for researchers. Over the years ^{12}C has been the object of study of several models, including Antisymmetric Molecular Dynamics[5], Fermion Molecular Dynamics[6], BEC-like cluster model[7], lattice EFT[8] and the Algebraic Cluster Model[9], all of them implementing or obtaining the cluster structure as a result. A more comprehensive review of many different models and their assumptions about clustering in nuclei can be found in [10–12].

Yet for its neighbors, a query arises when one questions to what extent the cluster structure persists with adding extra nucleons. This proceeding addresses part of this question, showing precursory results of longitudinal form factors for ^{12}C and ^{13}C and transverse form factors for ^{13}C in the framework of the Cluster Shell Model (CSM), where ^{13}C is seen as a ^{12}C core plus an extra neutron. ^{13}C can then be considered as a system with $\mathcal{D}'_{3h}$ symmetry, consisting of three α-particles in a triangular configuration plus an additional neutron moving in the deformed field generated by the cluster. The matrix elements of the form factor operators are then calculated, where a dominance of the core cluster structure is found for longitudinal form factors, exhibiting a very similar q dependence between ^{12}C and ^{13}C, corroborated by experimental data. Finally, we show some preliminary results for transverse form factors.

2. Cluster Shell Model

The Cluster Shell Model was introduced in [13–15] to describe nuclei composed of k α-particles plus additional nucleons, denoted as $k\alpha + x$ nuclei. It is similar in spirit to the Nilsson model [16]; the main difference lies in that the odd nucleon moves in the deformed field generated by the cluster core.

2.1. Cluster Potential

To obtain the single-particle energy levels, we define

$$H = T + V(\vec{r}) + V_{\rm so}(\vec{r}) + \frac{1}{2}(1+\tau_3)V_{\rm C}(\vec{r}) \; , \tag{1}$$

i.e. the sum of the kinetic energy, a central potential obtained by convoluting the density

$$\begin{aligned} \rho(\vec{r}) &= \left(\frac{\alpha}{\pi}\right)^{3/2} \sum_{i=1}^{k} \exp\left[-\alpha\left(\vec{r}-\vec{r}_i\right)^2\right] \\ &= \left(\frac{\alpha}{\pi}\right)^{3/2} {\rm e}^{-\alpha(r^2+\beta^2)}\, 4\pi \sum_{\lambda\mu} i_\lambda(2\alpha\beta r)\, Y_{\lambda\mu}(\theta,\phi) \sum_{i=1}^{k} Y^*_{\lambda\mu}(\theta_i,\phi_i) \; , \end{aligned} \tag{2}$$

with the interaction between the α-particle and the nucleon, a spin-orbit interaction, and a Coulomb potential for an odd proton. In equation (1), $\vec{r}_i = (r_i, \theta_i, \phi_i)$ represent the coordinates of the α-particles with respect to the center-of-mass of the cluster structure. The case of ^{13}C belongs to $k = 3$ with an equilateral triangular configuration, whose coordinates are given by $(\beta, \frac{\pi}{2}, 0)$, $(\beta, \frac{\pi}{2}, \frac{2\pi}{3})$ and $(\beta, \frac{\pi}{2}, \frac{4\pi}{3})$. The deformation parameter (from spherical symmetry) is β, which is the distance of each of the alpha particles to the center of mass of the cluster structure. The resulting potentials are

$$V(\vec{r}) = -V_0 \sum_{\lambda\mu} f_\lambda(r)\, Y_{\lambda\mu}(\theta,\phi) \sum_{i=1}^{k} Y^*_{\lambda\mu}(\theta_i,\phi_i)\, , \tag{3}$$

$$V_{so}(\vec{r}) = V_{0,so}\frac{1}{2}\left[\frac{1}{r}\frac{\partial V(\vec{r})}{\partial r}\left(\vec{s}\cdot\vec{l}\right) + \left(\vec{s}\cdot\vec{l}\right)\frac{1}{r}\frac{\partial V(\vec{r})}{\partial r}\right], \tag{4}$$

$$f_\lambda(r) = e^{-\alpha\left(r^2+\beta^2\right)} 4\pi i_\lambda\left(2\alpha\beta r\right). \tag{5}$$

2.2. $\mathcal{D}'_{3h}$ symmetry

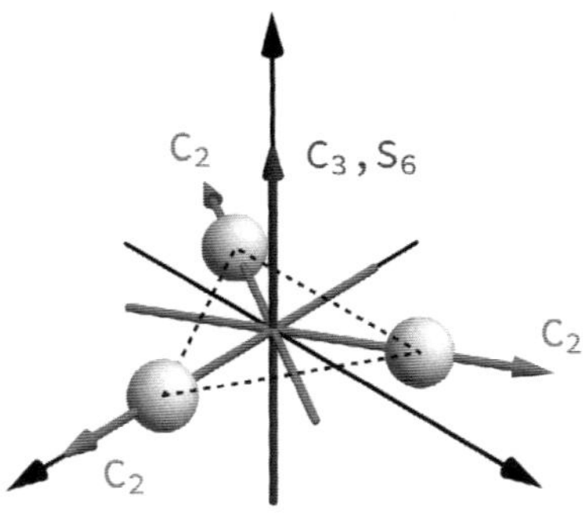

Figure 1. Schematic view of all principal axis and rotation operators for $\mathcal{D}'_{3h}$ symmetry.

To construct a complete wave function we need to understand in more detail the $\mathcal{D}'_{3h}$ symmetry. The objective is to construct a symmetry-adapted basis for $\mathcal{D}'_{3h}$ symmetry instead of a spherical basis[17; 18]. In the case of triangular symmetry the eigenstates of equation (1) can be classified according to the doubly degenerate spinor representations of the double point group $\mathcal{D}'_{3h}$ [19; 20]: $\Omega = E_{1/2}$, $E_{5/2}$ and $E_{3/2}$, or in the notation of Ref. [17] $\Omega = E^{(+)}_{1/2}$, $E^{(-)}_{1/2}$ and $E_{3/2}$, respectively. From similar work in molecular physics, to differentiate between the degenerate states, one finds all rotations of the structure of study; the case in question is shown in figure 1. Then improper rotations are used (in this case the S_6) to identify the intrinsic states of the doubly degenerate spinor representations. The resulting labels are $\gamma = \pm 5/2$ for $\Omega = E_{5/2}$, $\gamma = \pm 1/2$ for $E_{1/2}$ and $\gamma = \pm 3/2$ for $E_{3/2}$. The Hamiltonian of the CSM is solved in the body-fixed system, using the harmonic oscillator basis $|nljm\rangle$.

$$\phi_{\Omega\gamma} = \sum_{nljm} C^{\Omega\gamma}_{nljm} |nljm\rangle \tag{6}$$

The rotational states can be labeled by the angular momentum I, parity P, and its projection K on the symmetry axis, $|I^P K\rangle$. Both I and K are half integers. Then the allowed values of K^P for each one of the spinor representations, along with their γ labels are given by [17; 18]

$$\begin{array}{lcl} \Omega = E_{1/2},\ \gamma = \pm\frac{1}{2} & : & K^P = \pm\frac{1}{2}^+, \mp\frac{5}{2}^-, \pm\frac{7}{2}^-, \ldots \\ \Omega = E_{5/2},\ \gamma = \pm\frac{5}{2} & : & K^P = \mp\frac{1}{2}^-, \pm\frac{5}{2}^+, \mp\frac{7}{2}^+, \ldots \\ \Omega = E_{3/2},\ \gamma = \pm\frac{3}{2} & : & K^P = \pm\frac{3}{2}^+, \mp\frac{3}{2}^-, \mp\frac{9}{2}^+, \ldots \end{array} \tag{7}$$

with $I = |K|, |K|+1, \ldots$. A schematic view can be seen in figure 2. The complete wave function is defined then by [18]

$$|\Omega\gamma; I^P MK\rangle = \sqrt{\frac{2I+1}{16\pi^2}} \psi_v \left(1 + \mathcal{S}_i^{-1}\mathcal{S}_e\right) \phi_{\Omega\gamma} D^I_{MK}(\omega) \tag{8}$$

where ψ_v is the vibrational wave function, $\phi_{\Omega\gamma}$ the intrinsic wave function and $D^I_{MK}(\omega)$ the rotational wave function. The wave function is invariant under the transformation $\mathcal{S}_i^{-1}\mathcal{S}_e$ where the operator $\mathcal{S}$ is the product of a rotation about π followed by a parity transformation [18]. The operator $\mathcal{S}_i^{-1}$ acts on the intrinsic wave function and $\mathcal{S}_e$ on the rotational wave function.

2.3. Splitting of single-particle levels

Previous work on the splitting of single-particle levels is shown in [13; 21]. The splitting of the single-particle energy levels is seen in figure 3. Two simple observations that are illustrated in the figure are that at a value of $\beta = 0$ the spherical symmetry is recovered, and at max value, we have 3-fold degeneracy of the spherical symmetry. In other words, at the max value we obtain the view of a particle in a field formed by three separate (not bound in a nucleus) alpha particles, which is unlike the Nilsson model where in the case of max deformation one obtains the splitting for an "infinite cigar". From the elastic form factors, one obtains the value for β and then comes back to figures 2 and 3 and obtains the base rotational band proposed for ^{13}C. Foreshadowing the subsequent section, the value obtained for β is 1.71 fm, which in turn gives the ground state rotational band as $\Omega = E_{5/2}$, and the ground state with $I^P = \frac{1}{2}^-$.

3. Form Factors and comparison to experimental data

Using equation (8) we can now calculate the form factors for ^{13}C and compare them with available experimental data. We use the standard form factors operators, of which a full explanation is found in [23].

3.1. Longitudinal form factors

The charge distribution is taken to be equation (2), plus a point-like distribution for the extra nucleon

$$\rho(\vec{r}) = \frac{(Ze)_c}{3}\left(\frac{\alpha}{\pi}\right)^{3/2} \sum_{i=1}^{3} \exp\left[-\alpha\left(\vec{r} - \vec{r}_i\right)^2\right] + \tilde{e}\delta(\vec{r} - \vec{r}_{sp}), \tag{9}$$

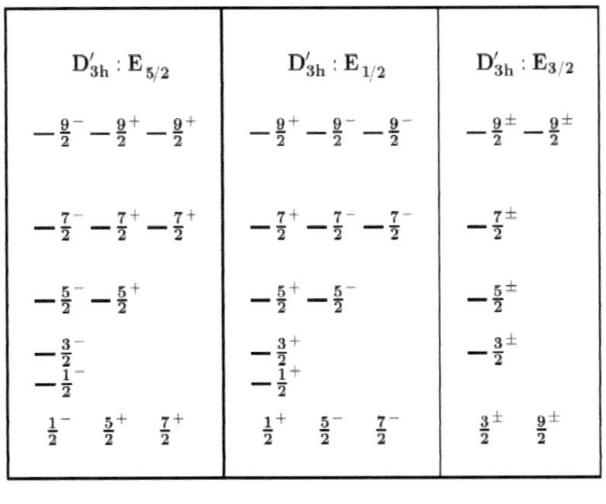

Figure 2. Schematic view of rotational structure with $\mathcal{D}'_{3h}$, proposed for ^{13}C. Imagine obtained from [22].

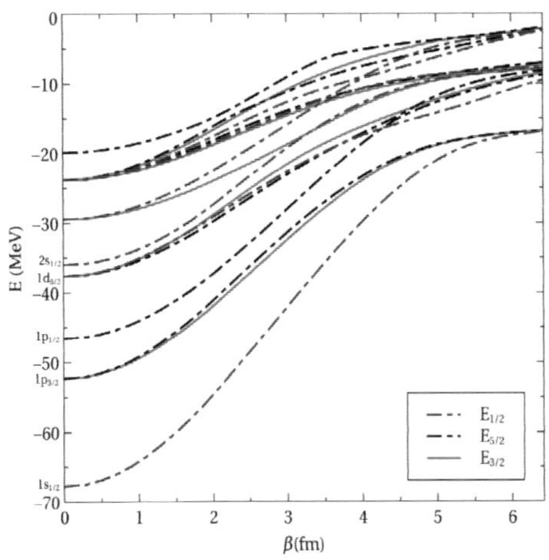

Figure 3. Single-particle energy level splitting from solving equation (1). The values used for equations (3) and (4) are V_0=32 MeV, α=0.0511 fm^{-2} and $V_{0,so}$=17 MeV fm^{-2}.

$(Ze)_\mathrm{c}$ is the electric charge of the 3α core nucleus, and $\tilde{e}$ the effective charge of the extra nucleon, which for the case of ^{13}C is taken to be zero. Making a Fourier transformation of equation (9), and then summing over final and averaged over initial states, we obtain the multipoles of the longitudinal (Coulomb) form factor;

$$F_{C\lambda}\left(q;\Omega'\gamma',I'K'\to\Omega\gamma,IK\right)=\frac{\sqrt{4\pi}}{(Ze)_{\mathrm{odd}}}\langle I',K',\lambda,K-K'\,|I,K\rangle\,\delta_{\Omega\Omega'}\delta_{\gamma\gamma'}G^{\mathrm{c}}_{vv'}\left(q\right), \tag{10}$$

where $(Ze)_{\mathrm{odd}}$ denotes the electric charge of the odd nucleus. For the vibrationally elastic case with $v=v'=0$ the collective part is given by

$$G^{\mathrm{c}}_{00}(q;\lambda,K-K')=(Ze)_\mathrm{c}\,j_\lambda(q\beta)\,\mathrm{e}^{-q^2/4\alpha}\,Y_{\lambda,K-K'}(\frac{\pi}{2},0)\,\delta_{K-K',3\kappa} \tag{11}$$

A final remark arises when calculating form factors with diagonal intrinsic states. In that case, branching ratios are found between different form factors with the same multipole.

$$\frac{|F_{C\lambda}(q;\Omega\gamma,I'K'\to\Omega\gamma,I_1K_1)|^2}{|F_{C\lambda}(q;\Omega\gamma,I'K'\to\Omega\gamma,I_2K_2)|^2}=\frac{\langle I',K',\lambda,K_1-K'|I_1,K_1\rangle^2}{\langle I',K',\lambda,K_2-K'|I_2,K_2\rangle^2}. \tag{12}$$

Finally one sees that the single particle element for Coulomb form factors doesn't give any contribution to the total value, thus we expect that for both ^{12}C and ^{13}C to have the same q dependence. The results are illustrated in figures 4 and 5.

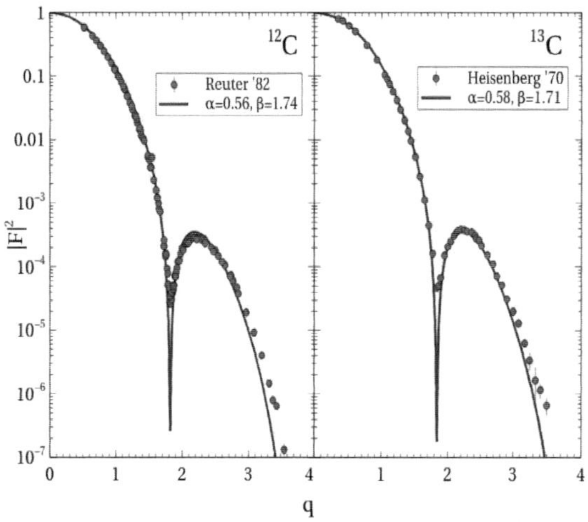

Figure 4. Comparison of elastic form factor between ^{12}C (left) and ^{13}C(right). Both graphs include available experimental data for both ^{12}C[24] and ^{13}C[25].

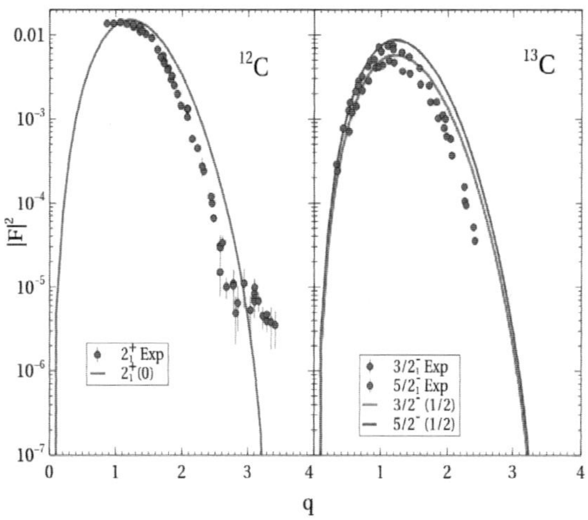

Figure 5. Comparison of C2 form factor between ^{12}C (left) and ^{13}C(right). Both graphs include available experimental data for both ^{12}C[26; 27] and ^{13}C[28].

3.2. Transverse form factors

Just as previously done for the Coulomb form factors, the transverse form factors are obtained from a Fourier transform of the current $\vec{J}$ and magnetization operator $\vec{\mu}$, with a little more algebra involved to obtain the final expressions. The main assumption we do from the CSM is that the only contributor to both is the extra nucleon

$$\vec{J} = 0 \tag{13}$$

$$\vec{\mu} = \delta\left(\vec{r_i} - \vec{x}\right)\mu_i \frac{\vec{\sigma}\left(i\right)}{2M_i}, \tag{14}$$

where μ_i is the magnetic moment (in nuclear magnetons) of the nucleon involved, in this case for the free neutron is $\mu_n = -1.91$ and M_i is the mass. The multipoles for the transverse form factor are then given by

$$\begin{aligned} F_{E\lambda/M\lambda}(q;\Omega'\gamma', I'K' =& \left\langle I', K', \lambda, K - K' | I, K \right\rangle \delta_{vv'} G^{\text{sp}}_{\Omega\gamma,\Omega'\gamma'}(q) \\ &+ P(-1)^{I+K}\left\langle I', K', \lambda, -K - K' | I, -K \right\rangle \delta_{vv'} H^{\text{sp}}_{\Omega\gamma,\Omega'\gamma'}(q), \end{aligned} \tag{15}$$

where $G^{\rm sp}_{\Omega\gamma,\Omega'\gamma'}(q)$ and $H^{\rm sp}_{\Omega\gamma,\Omega'\gamma'}(q)$ will have the form

$$G^{\rm sp}_{\Omega\gamma,\Omega'\gamma'}(q) = \sum_{nljm}\sum_{n'l'j'm'} C^{\Omega\gamma}_{nljm} C^{\Omega'\gamma'}_{n'l'j'm'} \left\langle nljm \left| \hat{T}^{\zeta}_{\lambda\mu}(q,\vec{r}) \right| n'l'j'm' \right\rangle \tag{16a}$$

$$H^{\rm sp}_{\Omega\gamma,\Omega'\gamma'}(q) = \sum_{nljm}\sum_{n'l'j'm'} C^{\Omega\gamma}_{nljm} C^{\Omega'\gamma'}_{n'l'j'm'} (-1)^{n+j+m} \left\langle nlj,-m \left| \hat{T}^{\zeta}_{\lambda\mu}(q,\vec{r}) \right| n'l'j'm' \right\rangle \tag{16b}$$

and $\hat{T}^{\zeta}_{\lambda\mu}(q,\vec{r})$ are the electric/magnectic transverse form factor operators

$$\begin{aligned}
\hat{T}^{E}_{\lambda\mu}(q,\vec{r}) &= q\frac{\mu_i}{2M_N}\vec{M}^{\mu}_{\lambda,\lambda}(\vec{r}_i)\cdot\vec{\sigma}(i) \\
\hat{T}^{M}_{\lambda\mu}(q,\vec{r}) &= iq\frac{\mu_i}{2M_N}\left(-\sqrt{\frac{\lambda}{2\lambda+1}}\vec{M}^{\mu}_{\lambda,\lambda+1}(\vec{r}_i) + \sqrt{\frac{\lambda+1}{2\lambda+1}}\vec{M}^{\mu}_{\lambda,\lambda-1}(\vec{r}_i)\right)\cdot\vec{\sigma}(i) \\
\vec{M}^{\mu}_{\lambda_1,\lambda_2}(\vec{r}) &= j_{\lambda_2}(qr)\vec{Y}^{\mu}_{\lambda_1\lambda_2 1}(\theta,\phi).
\end{aligned} \tag{17}$$

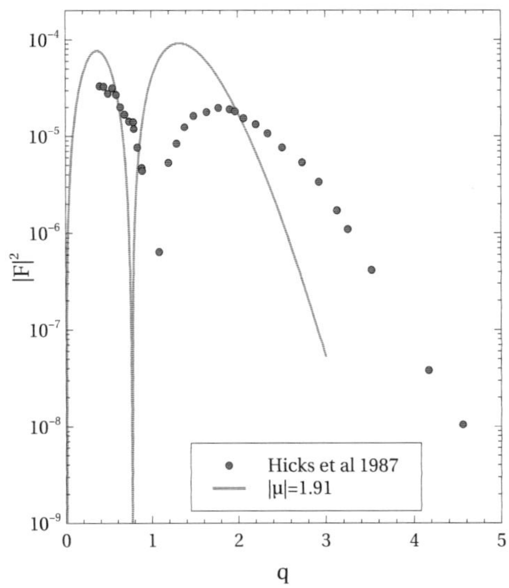

Figure 6. Comparison of M1 elastic form factor between experimental data [28] and CSM calculation.

The results are showed in figure 6.

4. Summary and Conclusions

From calculations, it is seen that for longitudinal form factors in the base rotational band of ^{13}C, the cluster core structure is dominant. In those cases, we expected that Coulomb form factors for both ^{12}C and ^{13}C have the same q dependence. From figures 4 and 5 we see that the experimental data seems to support that idea, differing only slightly in some aspects. Also, it shows for ^{13}C that longitudinal form factors follow a branching ratio rule like the one in equation (12). This is of great importance since it indicates that some properties of the cluster structure in ^{12}C are still present in ^{13}C.

Unfortunately, such a positive outcome was not the case for the transverse form factors. From figure 6 we are not able to reproduce available experimental data. Only global functional properties are reproduced of [28]. It is most likely that our assumptions in equations (13) and (14) are incorrect and must then include a contribution of the cluster core. A possible solution may be found in [29], but it would have to be adapted for the CSM.

To reiterate, the CSM as is now is a good approximation to calculate longitudinal form factors for ^{13}C, reproducing accurately the available experimental data. Meanwhile, in the case of transverse form factors, the model is still lacking, due most likely to not considering the contribution of the cluster structure to the transverse form factors.

Acknowledgments

This work was supported in part by research grants IN101320, IG101423 from PAPIIT-DGAPA and 784896 from CONACYT.

References

[1] Wheeler J A 1937 *Phys. Rev.* **52** 1083–1106

[2] Hafstad L R and Teller E 1938 *Phys. Rev.* **54** 681–692

[3] Brink D 1965 *Int. School of Physics Enrico Fermi (Varenna, 1965)* 247

[4] Brink D, Friedrich H, Weiguny A and Wong C 1970 *Phys. Lett. B* **33** 143–146

[5] Kanada-En'yo Y 2007 *Prog. Theor. Phys.* **117** 655–680

[6] Chernykh M, Feldmeier H, Neff T, von Neumann-Cosel P and Richter A 2007 *Phys. Rev. Lett.* **98** 032501

[7] Funaki Y, Horiuchi H, von Oertzen W, Röpke G, Schuck P, Tohsaki A and Yamada T 2009 *Phys. Rev.* C **80** 064326

[8] Epelbaum E, Krebs H, Lee D and Meißner U G 2011 *Phys. Rev. Lett.* **106** 192501

[9] Bijker R and Iachello F 2002 *Ann. Phys.* **298** 334–360

[10] Freer M and Fynbo H 2014 *Prog. Part. Nucl. Phys.* **78** 1–23

[11] Schuck P, Funaki Y, Horiuchi H, Röpke G, Tohsaki A and Yamada T 2016 *Phys. Scr.* **91** 123001

[12] Freer M, Horiuchi H, Kanada-En'yo Y, Lee D and Meißner U G 2018 *Rev. Mod. Phys.* **90** 035004

[13] Della Rocca V, Bijker R and Iachello F 2017 *Nucl. Phys.* A **966** 158–184

[14] Della Rocca V and Iachello F 2018 *Nucl. Phys.* A **973** 1–32

[15] Bijker R and Iachello F 2020 *Prog. Part. Nucl. Phys.* **110** 103735

[16] Nilsson S G 1955 *Kong. Dan. Vid. Sel. Mat. Fys. Med.* **29** 1–69

[17] Bijker R and Iachello F 2019 *Phys. Rev. Lett.* **122** 162501

[18] Santana Valdés A H and Bijker R 2020 *Eur. Phys. J. ST* **229** 2353–2366

[19] Herzberg G 1991 *Electronic Spectra and Electronic Structure of Polyatomic Molecules* (*Molecular spectra and molecular structure* vol II) (Malabar, Fla: R.E. Krieger Pub. Co.) ISBN 0894642685

[20] Koster G F, Dimmock J O, Wheeler R G and Statz H 1963 *Properties of the thirty-two point groups* vol 24 (MIT press Cambridge, MA)

[21] Santana-Valdés A and Bijker R 2018 *J. Phys. Conf. Ser.* **1078** 012019

[22] Bijker R and Santana-Valdés A 2020 *Journal of Physics: Conference Series* **1643** 012113

[23] de Forest T and Walecka J 1966 *Adv. Phys.* **15** 1–109

[24] Reuter W, Fricke G, Merle K and Miska H 1982 *Phys. Rev.* C **26** 806–818

[25] Heisenberg J, McCarthy J and Sick I 1970 *Nucl. Phys.* A **157** 435–448

[26] Crannell H L and Griffy T A 1964 *Phys. Rev.* **136** B1580–B1584

[27] Crannell H 1966 *Phys. Rev.* **148** 1107–1118

[28] Millener D J, Sober D I, Crannell H, O'Brien J T, Fagg L W, Kowalski S, Williamson C F and Lapikás L 1989 *Phys. Rev.* C **39** 14–46

[29] Delorme J, Figureau A and Guichon P 1981 *Physics Letters B* **99** 187–190

Journal of Physics: Conference Series **2619** (2023) 012016 doi:10.1088/1742-6596/2619/1/012016

A multi–channel study of the ^{20}Ne + ^{130}Te system within the NUMEN project

V Soukeras[1,2], L Acosta[3], C Agodi[2], I Boztosun[4], G A Brischetto[1,2], D Calvo[5], F Cappuzzello[1,2], D Carbone[2], M Cavallaro[2], E R Chávez Lomelí[3], I Ciraldo[1,2], F Delaunay[1,2,6], P Finocchiaro[2], M Fisichella[2], A Hacisalihoglu[2,7], G Lanzalone[2,8], R Linares[9], J R B Oliveira[10], A Pakou[11], L Pandola[2], H Petrascu[12], F Pinna[5,13], O Sgouros[2], S O Solakci[4], G Souliotis[14], A Spatafora[1,2], D Torresi[2], S Tudisco[2], A Yildirim[4], V A B Zagatto[10], for the NUMEN collaboration

[1] Dipartimento di Fisica e Astronomia "Ettore Majorana", Università di Catania, Catania, Italy
[2] INFN – Laboratori Nazionali del Sud, Catania, Italy
[3] Instituto de Física, Universidad Nacional Autónoma de México, Mexico City, Mexico
[4] Department of Physics, Akdeniz University, Antalya, Turkey
[5] INFN – Sezione di Torino, Torino, Italy
[6] LPC Caen, Normandie Université, ENSICAEN, UNICAEN, CNRS/IN2P3, Caen, France
[7] Institute of Natural Science, Karadeniz Teknik Universitesi, Trabzon, Turkey
[8] Facoltà di Ingegneria e Achitettura, Università di Enna "Kore", Enna, Italy
[9] Instituto de Fisica, Universidade Federal Fluminense, Niteroi, Brazil
[10] Instituto de Fisica, Universidade de Sao Paulo, Sao Paulo, Brazil
[11] Department of Physics, University of Ioannina and Hellenic Institute of Nuclear Physics, Ioannina, Greece
[12] IFIN-HH, Bucarest, Romania
[13] DISAT – Politecnico di Torino, Torino, Italy
[14] Department of Chemistry, University of Athens and Hellenic Institute of Nuclear Physics, Athens, Greece

E-mail: `vasileios.soukeras@lns.infn.it` (V. Soukeras)

Abstract. The NUMEN project aims to measure specific reaction cross sections to provide experimentally driven information about nuclear matrix elements of interest in the context of neutrinoless double beta decay ($0\nu\beta\beta$). In particular, it was proposed to use heavy – ion induced double charge exchange reactions as tools towards the determination of information on the nuclear matrix elements of $0\nu\beta\beta$, strongly motivated by a number of similarities between the two processes. To this extent, the ^{20}Ne + ^{130}Te system was experimentally investigated in a multi-channel approach by measuring the complete net of reactions channels, namely double charge exchange, single charge exchange, elastic and inelastic scattering, one – and two – nucleon transfer reactions, characterized by the same initial projectile and target nuclei. The goal of the study is to fully characterize the properties of the nuclear wavefunctions entering in the $0\nu\beta\beta$ decay nuclear matrix elements. The experimental setup, the data reduction and some of the obtained results for the ^{20}Ne + ^{130}Te system will be presented and discussed.

Journal of Physics: Conference Series **2619** (2023) 012016 doi:10.1088/1742-6596/2619/1/012016

1. Introduction

Neutrinoless double beta decay ($0\nu\beta\beta$) is one of the "hot" subject in physics research nowadays [1, 2]. The process, forbidden in the Standard Model framework, has not been experimentally observed yet and only experimental lower limits on the $0\nu\beta\beta$ half-time are available for some isotopes [3, 4, 5, 6, 7]. In this extremely rare yet elusive process, a nucleus with mass and atomic numbers (A,Z) is transformed into a new one with mass and atomic numbers (A,Z+2) emitting also two electrons or to (A,Z-2) emitting two positrons. Since no neutrinos or antineutrinos are emitted, the aforementioned processes violate the lepton number conservation law. The decay rate of the $0\nu\beta\beta$ together with a precise determination of its nuclear matrix elements (NMEs) provide a unique tool to access the effective neutrino mass. The measurement of the half-life of $0\nu\beta\beta$ is a challenging aspect but, the uncertainties met in the NMEs determination are also large, approaching a factor of 3-4 introducing an uncertainty of an order of magnitude in the effective neutrino mass [2, 8].

To this extent, the NUMEN (NUclear Matrix Elements for Neutrinoless double beta decay) project [9, 10] aims to access information on the $0\nu\beta\beta$ NMEs through the study of the heavy-ion (HI) induced double charge exchange (DCE) reactions on all $\beta\beta$ decay candidate targets. This is strongly motivated by a number of similarities between the two processes taking also into account that the NMEs of the two processes appeared to be in linear correlation [11, 12, 13]. A precise determination of the NMEs of the surrogate process required not only the measurement of the DCE reaction but also, all the reactions channels characterized by the same initial projectile and target nuclei. Therefore, NUMEN aims to experimentally investigate the complete net of reaction channels including the:

- Double Charge Exchange reaction (DCE),
- Single Charge Exchange reaction (SCE),
- one-proton transfer reaction,
- one-neutron transfer reaction,
- two-proton transfer reaction,
- two-neutron transfer reaction,
- elastic scattering, and
- inelastic scattering channel.

An illustrative example of the reaction paths for the system discussed in this article is presented in Fig. 1

Several HI systems are currently investigated in this phase of the NUMEN project [10, 14, 15, 16, 17, 18, 19, 20, 21, 22, 23, 24, 25, 26, 27]. The heaviest of the targets explored so far in NUMEN is the ^{130}Te [18, 21, 28]. CUORE experimental campaign [3] is also searching to measure the $0\nu\beta\beta$ half-life for this nucleus providing so far experimental limit for the $0\nu\beta\beta$ half-life. Another experimental campaign in construction, SNO+ [29], also aims to explore the ^{130}Te as a $0\nu\beta\beta$ candidate. The first experiment related to the study of the ^{20}Ne + ^{130}Te system within the NUMEN and NURE projects [9, 30] will be presented in the present article, providing a first input to a forthcoming exploration of the system in the next phase of the NUMEN project.

2. Experimental details

The relevant experiment was visualized at the MAGNEX facility[31] of the Istituto Nazionale di Fisica Nucleare - Laboratori Nazionali del Sud (INFN-LNS) in Catania, Italy where the NUMEN experimental campaign takes place. A $^{20}Ne^{10+}$ beam was accelerated by the K800 Superconducting Cyclotron at 306 MeV (15.3 MeV/nucleon) incident energy and impinged on a $\sim 250\ \mu g/cm^2$ ^{130}Te target evaporated onto a thin carbon foil with a thickness of $\sim 40\ \mu g/cm^2$. Two different post - stripper materials, C_3H_6 and carbon (each one in a different set of runs),

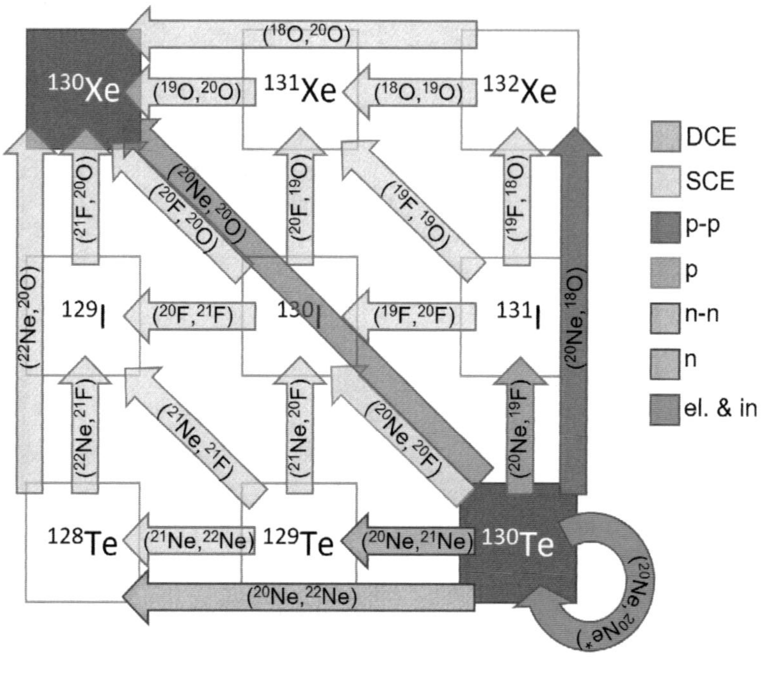

Figure 1. Network of possible nuclear reaction routes connecting initial and final states in the $^{20}Ne+^{130}Te$ DCE reaction. Colour labels are indicated in the legend.

were located downstream the target position in order to minimize the amount of $^{20}Ne^{8+}$ and $^{20}Ne^{9+}$ events [28].

The various ejectiles were momentum analyzed by the MAGNEX magnetic spectrometer while the different ions were detected by the MAGNEX Focal Plane Detector (FPD) [32, 33]. The MAGNEX FPD consists of a gas tracker followed by a wall of 60 single silicon detectors without any intermediate foils but with only an entrance mylar window. The gas tracker includes six sections each one having at the top a proportional wire (DC). A set of 224 induction pads is mounted above each DC wire allowing the measurement of the horizontal position (X_f) and angle (θ_f). The present experiment was the first one on which the new configuration with the six DCs [32] was used. Furthermore, the electron drift time measurements inside the gas allow for the determination of the vertical position (Y_f) and angle (ϕ_f). The energy loss (ΔE_{tot}) of the ions inside the gas is measured by the DCs wires while, the residual energy E_{resid} is measured by the silicon detectors. Finally, the angular and energy coverage of MAGNEX is presented in Table 1 for the measurement of the different reaction channels.

3. Data reduction

The particle identification (PID) was performed as described in Ref. [34]. In particular, the first step of the PID for the reaction channels of our interest is the separation based on the ions atomic number (Z) by means of the standard ΔE–E technique. A typical ΔE–E spectrum is shown in Fig. 2 for a single silicon detector. After selecting the Z of interest, the mass identification is feasible via the X_f–E_{resid} (horizontal position versus residual energy measured by a single silicon detector) spectra through the relation:

$$X_f \propto \frac{\sqrt{m}}{q}\sqrt{E_{resid}}, \tag{1}$$

Table 1. Angular and energy coverage of MAGNEX in the measurement of the different reaction channels. For each reaction channel of interest (first column), the detected ion and the angle(s) of the MAGNEX optical axis are presented in the second and third column, respectively. The covered angular range (fourth column) for a specific E_x range (fifth column) is also presented.

Reaction	Detected ion	θ_{opt} (°)	θ_{lab} (°)	E_x (MeV)	See also
DCE	^{20}O	-3	0 to 9.5	0 to 10	Ref. [21]
			0 to 6	10 to 25	Ref. [21]
SCE	^{20}F	10	7 to 15	0 to 35	
1p-transfer	^{19}F	10	8 to 15	0 to 28	
1n-transfer	^{21}Ne	10	8 to 15	0 to 4	
2p-transfer	^{18}O	-3	0 to 7	0 to 8	
2n-transfer	^{22}Ne	10	8 to 15	0 to 14	
(quasi-)elastic	^{20}Ne	8, 13, 20	3 to 26		Ref. [18]
inelastic	^{20}Ne	13, 20	13 to 23	$\sim$ 1.6	Ref. [18]
			14 to 19	$\sim$ 2.5	Ref. [18]

where m and q are the particle mass and charge state, respectively. In this way, the ions with a different $\frac{\sqrt{m}}{q}$ ratio, appear at different loci at the X_f–E_{resid} representation as it is shown in Figs. 3,4 For all the cases of interest, both atomic number and mass separation is excellent. This is essential particularly for reaction channels with extremely tiny cross section, such as the DCE one.

Having identified the reaction channels of interest, a software high – order (10th) ray reconstruction is applied to the data [35] and, the excitation energy spectra are obtained. Subsequently, the absolute cross sections are deduced for each of the reaction channels of interest.

4. Results and perspectives

The excitation energy spectrum for the ^{130}Te(^{20}Ne,^{20}O)^{130}Xe DCE reaction, shown in Fig. 5 [21], allows for the determination of the DCE cross section for the $^{130}\mathrm{Te}_{g.s.} \to {}^{130}\mathrm{Xe}_{g.s.}$ transition, the best estimate of which is 13 nb (between 3 and 18 nb in a 95% confidence level) in the angular range $0.0° \leq \theta_{lab} \leq 9.5°$ and in the energy range -1 MeV $\leq E_x \leq$ 1 MeV [21]. Except the transition to the $^{20}\mathrm{O}_{g.s.}(0^+)$ + $^{130}\mathrm{Xe}_{g.s.}(0^+)$, a contribution due to the transition to the $^{20}\mathrm{O}_{g.s.}(0^+)$ + $^{130}\mathrm{Xe}_{0.536}(2^+)$ may be expected due to the finite experimental energy resolution which is estimated at $\sim$ 0.5 MeV from similar experiments in MAGNEX facility [19, 20]. It should be underlined the absence of any spurious events in the negative E_x region while, any background due to to target backing and post-stripper material is expected at $E_x >$ 33 MeV due to kinematics. The measurement provides, for the first time, an estimation of the tiny DCE cross section for the transition under study. It should be noted that the ^{130}Te is the heaviest of the targets under investigation at the second phase of the NUMEN project and one of the heaviest targets to be investigated in the fourth phase of the project. The latter is launched after the upgrade of the INFN - LNS Superconducting Cyclotron and the MAGNEX FPD, more details of which can be found in Refs. [36, 37, 39, 40].

Except the experimental investigation of the DCE reaction presented in Ref. [21], the data analysis of the quasi-elastic and inelastic scattering channels is already completed and reported in Ref. [18]. The quasi-elastic and inelastic scattering cross-section angular distributions were

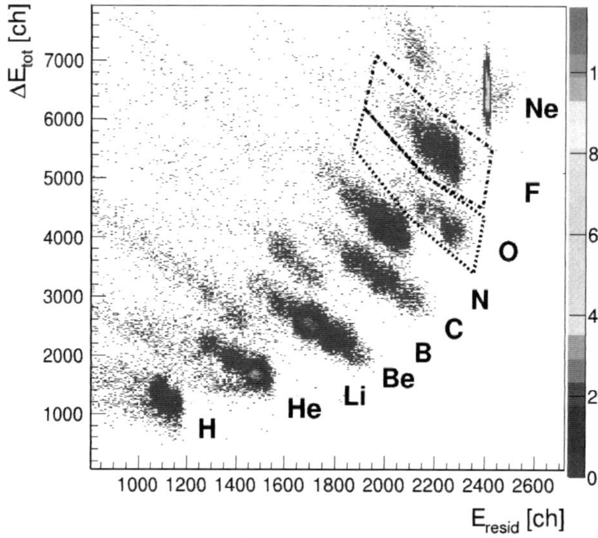

Figure 2. Typical ΔE_{tot} - E_{resid} spectrum for a single silicon detector of the FPD. Graphical selections of oxygen and fluorine ions are depicted with dashed and dot-dashed contour respectively.

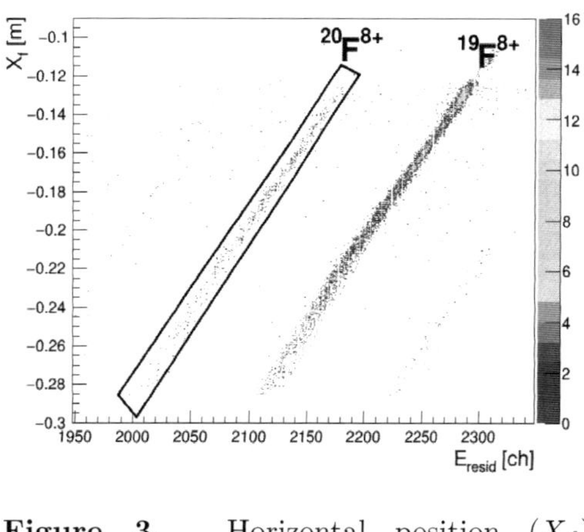

Figure 3. Horizontal position (X_f) versus residual energy (E_{resid}) spectrum after applying the graphical cut on fluorine ions of Fig. 2. The graphical selection of $^{20}F^{8+}$ (SCE) is illustrated with the solid line. See text for more details.

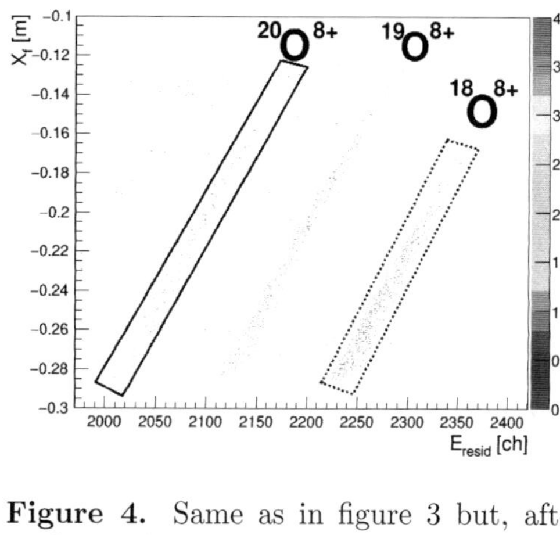

Figure 4. Same as in figure 3 but, after applying the graphical cut on oxygen ions of Fig. 2. The graphical selections of $^{20}O^{8+}$ (DCE) and $^{18}O^{8+}$ (2p-transfer) ions are illustrated with the solid and dashed line respectively.

compared to theoretical calculations within the optical model, one-step distorted wave Born approximation and coupled-channels frameworks, presenting in overall a very good agreement among them [18]. Finally, the data analysis for the rest of the reaction channels, i.e. the SCE,

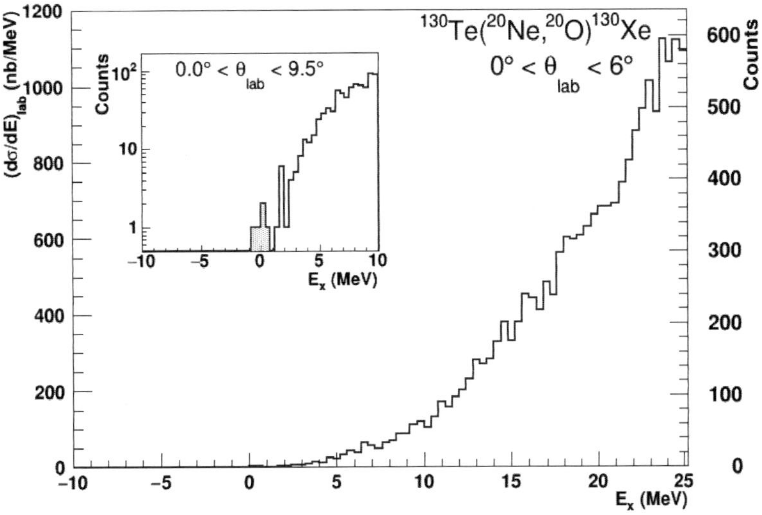

Figure 5. Excitation energy spectrum for the ^{130}Te(^{20}Ne,^{20}O)^{130}Xe DCE reaction at 306 MeV incident energy and $0.0° \leq \theta_{lab} \leq 6.0°$ angular range. A zoomed view for $E_x < 10$ MeV and full angular range $0.0° \leq \theta_{lab} \leq 9.5°$ is shown in the inset. The $^{130}Te_{g.s.} \rightarrow {}^{130}Xe_{g.s.}$ region (-1 MeV $\leq E_x \leq$ 1 MeV) is indicated with the red hatched area. Figure from Ref. [21].

one-nucleon and two-nucleon transfer, is in progress.

It should be pointed out that two successive SCE reactions, multi-nucleon transfer processes or a combination among them may -in principle- lead to the production of ^{20}O ejectiles. The measurement of the SCE channel and the single- and two-nucleon transfer reactions may provide the appropriate constraints on the estimated contribution of these competitive processes to the DCE cross section. As regards the multi-nucleon transfer, a recent theoretical work for the ^{20}Ne+^{116}Cd system at the same energy with the present study pointed to a negligible contribution of such process to the overall DCE cross-section [38].

The advent of the upgrade of the LNS facilities will allow for a deeper investigation of the ^{20}Ne + ^{130}Te system at different incident energies while, an investigation of the ^{130}Xe nucleus is also planned (symmetric processes to the ones shown in Fig. 1). Into this context, the development of a gas or implanted ^{130}Xe target is necessary and a relevant research is currently in progress [39, 41, 42, 43].

Acknowledgments

This project has received funding from the European Research Council (ERC) under the European Union's Horizon 2020 research and innovation programme (NURE - Grant agreement No. 714625).

References

[1] Dolinski D J, Poon A W P and Rodejohann W 2019 *Annual Reviews of Nuclear and Particle Science* **69** 219
[2] Ejiri H, Suhonen J and Zuber K 2019 *Physics Reports* **797** 1
[3] Adams D Q et al 2022 *Nature* **604** 53
[4] Anton G et al 2019 *Physical Review Letters* **123** 161802
[5] Abe S et al 2023 *Physical Review Letters* **130** 051801

[6] Agostini M et al 2020 *Physical Review Letters* **125** 252502
[7] Cattadori C M and Salamida F 2021 *Universe* **7** 314
[8] Agostini M, Benato G, Detwiler J A, Menéndez J and Vissani F 2022 *arxiv:2202.01787*
[9] Cappuzzello F, Agodi C, Cavallaro M et al 2018 *European Physical Journal* A **54** 72
[10] Cappuzzello F, Lenske H, Cavallaro M et al 2023 *Progress in Particle and Nuclear Physics* **128** 103999
[11] Shimizu N, Menéndez J and Yako K 2018 *Physical Review Letters* **120** 142502
[12] Santopinto E, García-Tecocoatzi H, Magaña Vsevolodovna R I and Ferretti J 2018 *Physical Review C* **98** 061601(R)
[13] Lenske H, Cappuzzello F, Cavallaro M and Colonna M 2019 *Progress in Particle and Nuclear Physics* **109** 103716
[14] Cappuzzello F, Cavallaro M, Agodi C et al 2015 *European Physical Journal* A **51** 145
[15] Calabrese S, Cappuzzello F, Carbone D et al 2018 *Acta Physica Polonica B* **49** 275
[16] Spatafora A, Cappuzzello F, Carbone D et al 2019 *Physical Review C* **100** 034620
[17] Carbone D, Ferreira J L, Calabrese S et al 2020 *Physical Review C* **102** 044606
[18] Carbone D, Linares R, Amador-Venenzuela P et al 2021 *Universe* **7** 58
[19] Cavallaro M, Bellone J I, Calabrese S et al 2021 *Frontiers in Astronomy and Space Sciences* **8** 659815
[20] Ferreira J L, Carbone D, Cavallaro M et al 2021 *Physical Review C* **103** 054604
[21] Soukeras V, Cappuzzello F, Carbone D et al 2021 *Results in Physics* **28** 104691
[22] Sgouros O, Cavallaro M, Carbone D et al 2021 *Physical Review C* **104** 034617
[23] La Fauci L, Spatafora A, Cappuzzello F et al 2021 *Physical Review C* **104** 054610
[24] Calabrese S, Cavallaro M, Carbone D et al 2021 *Physical Review C* **104** 064609
[25] Burrello S, Calabrese S, Cappuzzello F, Carbone D et al 2022 *Physical Review C* **105** 024616
[26] Ciraldo I, Cappuzzello F, Carbone D et al 2022 *Physical Review C* **105** 044607
[27] Spatafora A, Carbone D, Cappuzzello F et al 2023 *Physical Review C* **107** 024605
[28] Cavallaro M, Santagati G, Cappuzzello F et al 2019 *Results in Physics* **13** 102191
[29] Albanese V, Alves R, Anderson M R et al 2021 *Journal of Instrumentation* **16** P08059
[30] Cavallaro M et al 2017 *Proceedings of Science* **302** BORMIO2017_015
[31] Cappuzzello F, Agodi C, Carbone D and Cavallaro M 2016 *European Physical Journal* A **52** 167
[32] Torresi D, Sgouros O, Soukeras V et al 2021 *Nuclear Instruments and Methods in Physics Research Section* A **989** 164918
[33] Cavallaro M, Cappuzzello F, Carbone D et al 2012 *European Physical Journal* A **48** 59
[34] Cappuzzello F, Cavallaro M, Cunsolo A et al 2010 *Nuclear Instruments and Methods in Physics Research Section* A **621** 419
[35] Cappuzzello F, Carbone D and Cavallaro M 2011 *Nuclear Instruments and Methods in Physics Research Section* A **638** 74
[36] Agodi C, Russo A D, Calabretta L et al 2021 *Universe* **7** 72
[37] Cappuzzello F, Acosta L, Agodi C et al 2021 *Frontiers in Astronomy and Space Sciences* **8** 668587
[38] Ferreira J L, Lubian J, Cappuzzello F et al 2022 *Physical Review C* **105** 014630
[39] Cappuzzello F, Agodi C Calabretta L et al 2021 *International Journal of Modern Physics A* **36** 2130018
[40] Finocchiaro P, Acosta L, Agodi C et al 2020 *Universe* **6** 129
[41] Agodi C, Bruno E, Cappuzzello F et al *project for the "Feasibility study of implantation of Xe in High Oriented Pyrolithic Graphite substrate"* by using the ion implanter of the Department of Physics and Astronomy of the University of Catania
[42] Sandoval-Hipolito S, Huerta A, Mas-Ruiz J, Andrade E and Chavez E 2022 *poster presented in the 44th Symposium on Nuclear Physics (SNP'23), Cocoyoc, Mexico*
[43] Esquivel-Carrillo C, Valdez A O, Andrade E et al 2022 *poster presented in the 44th Symposium on Nuclear Physics (SNP'23), Cocoyoc, Mexico*

Journal of Physics: Conference Series **2619** (2023) 012017 doi:10.1088/1742-6596/2619/1/012017

Optimization of BeO formation parameters for the measurement of ^{10}Be by AMS.

L. R. Romero[1], C. J. Ramírez-Martínez[1], C. G. Méndez-García[2], S. Padilla[3], C. Solís[4] and H. Huerta[4].

liviaromero@estudiantes.fisica.unam.mx

[1] Facultad de Ciencias, Universidad Nacional Autónoma de México (UNAM), Mexico City, Mexico.
[2] INVESTIGADORA POR MÉXICO CONACyT - Instituto de Física, Universidad Nacional Autónoma de México (UNAM), Mexico City, Mexico.
[3] Facultad de Ciencias Experimentales, Universidad de Huelva, España.
[4] Instituto de Física, Universidad Nacional Autónoma de México (UNAM), Mexico City, Mexico.

Abstract. ^{10}Be has been widely studied over the last few years. The sensitivity offered by the Accelerator Mass Spectrometry technique (AMS) expands the efficient detection of ^{10}Be by 4-5 orders of magnitude. The chemical species most used in the measurement of ^{10}Be by AMS is beryllium oxide (BeO), which has been shown to offer the best conditions for measurements. In this work, several parameters in the BeO formation have been analyzed. The results show the efficiency of the protocol used but suggest that using Ag instead of Nb as a conductor could improve the measurement conditions.

1. Introduction

^{10}Be is a radioisotope usually produced by the interaction of cosmic rays with the components of the atmosphere, which are ^{14}N and ^{16}O. It is also produced, in smaller amounts, in the rocks by spallation reactions[1]. Meteoric ^{10}Be occurs mainly in the lower stratosphere and upper troposphere, with residence times of several years and about two weeks, respectively[2]. During these times, various atmospheric mixing occurs, such as stratosphere-troposphere exchange and atmospheric deposition transport that can influence surface ^{10}Be concentrations and deposition fluxes. ^{10}Be is attached to aerosols and is subsequently deposited on the earth's surface[1]. Once deposited on the surface, ^{10}Be adsorbs on particles with a pH greater than 6.

Meteoric ^{10}Be has been widely applied to different problems in recent decades. The study in rainwater[3-6], surface sediments[7-9], ice cores[10-11] and particulate matter[12-16] provides important information related to production and distribution in the atmosphere. It is also used to study atmospheric mixing, stratosphere-troposphere, as well as exchange and seasonal variations, which can provide the transport time from production to surface deposition of this radioisotope.

The high sensitivity of AMS allows precise measurements of ^{10}Be concentrations in samples, in some cases, it is even the only possible way to quantify it. ^{10}Be half - life ($t_{1/2}$=1.36±0.07 Ma)[2] making it a perfect candidate for dating samples, landscape reconstruction, and the study of erosion surfaces. It is considered an environmental tracer[3]. The AMS facility at the National Laboratory of Accelerator Mass Spectrometry (LEMA), is based on a 1 MV High Voltage Europe Engineering isotope separator. Recently, new lines of research began to be opened for the analysis of ^{10}Be in environmental samples with different applications.

The ^{10}Be extraction protocols used at LEMA look for the best conditions for measuring efficiency and lowest background contributions given its low natural abundance. The optimization is essential since it maximizes current and stability, with which better statistics and higher precision in the $^{10}Be/^{9}Be$ ratio are obtained, resulting in more accurate ^{10}Be concentrations[1]. In this work, some parameters from the initial protocol were analyzed to have an optimized protocol. The optimal parameters were found in beam transmission, currents, and stability.

2. Methodology

The general protocol used for the extraction of ^{10}Be in natural samples is shown in Figure 1, its purpose is to extract Be from the samples and precipitate it as BeO, which can then be used as a target for the AMS measurement.

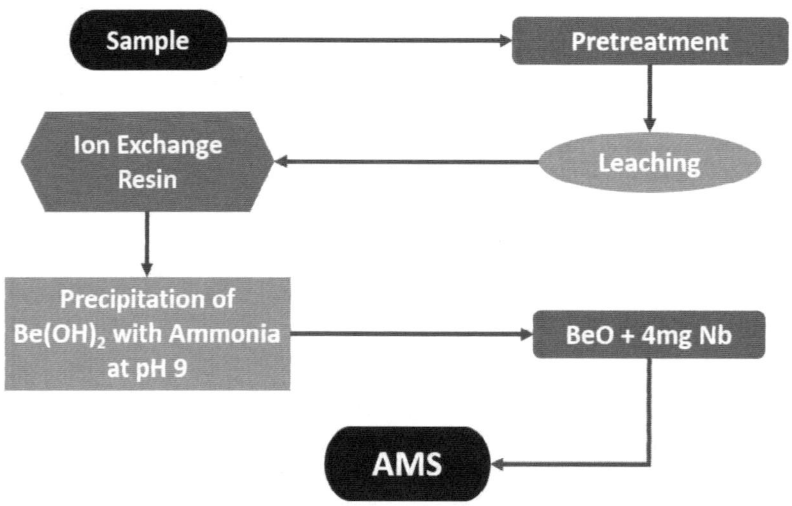

Figure 1. General diagram of measurement of ^{10}Be.

The protocol begins by adding ^{9}Be in the samples because AMS requires high currents of the stable isotope, and the presence of ^{9}Be is null in natural samples. Then, the samples are then pre-treated (depending on their origin) and leached with different acids to make ^{10}Be available. After leaching, ^{10}Be is separated and purified using different ion exchange resins. Firstly, the samples are passed through an anion exchange resin Bio-Rad AG1-X8, to remove Fe and Mg, after ^{10}Be is separated using cation exchange resin AG50W-X8.

Journal of Physics: Conference Series **2619** (2023) 012017 doi:10.1088/1742-6596/2619/1/012017

The result of the previous step is the purified Be fraction dissolved in a slightly acidic dilution. From this dilution, the Be is precipitated as $Be(OH)_2$ by the addition of ammonia to pH = 9. The $Be(OH)_2$ is freeze-dried for 24 h and subsequently calcined in a furnace at 1000°C for 1 h, whereupon BeO is obtained. This BeO is mixed with 4 mg which serves as a thermoelectric conductor to promote the formation of negative ions. Finally, the BeO + Nb mixture is pressed on aluminium cathodes and measured by AMS for 1h. Measured isotopic ratios $^{10}Be/^{9}Be$ were normalized to a standard sample with $^{10}Be/ ^{9}Be = 2.709\times10^{-11}$ [18].

For the optimization of the BeO formation parameters, the intensity and stability of the beam currents were analyzed. In this case, ^{10}Be targets were prepared from 250 µL of the standard ^{9}Be dilution ($Be_4O(C_2H_3O_2)_6$ standard, 1000 mg/L for ICP-MS, Merck). Each variation in the different parameters was done in triplicate.

The first parameter analyzed was pH, the pH value established in the protocol is 9, in this study, the current variations were analyzed with BeO formed from $Be(OH)_2$ precipitated with ammonia at pH 7, 9, and 12. The second parameter was the calcination temperature; temperatures of 900°C, 1000°C, and 1100°C were examined. Lastly, the third and fourth parameters were the variations with different metallic matrices and their ratios.

The tuning was performed at +1 charge state both at the accelerator output and after the passive absorber (75 nm silicon nitride window)[19-20]. The behaviour and evolution of the beam currents of the different variations were measured in the high-energy zone every five minutes.

3. Results

The results of variation of pH, calcination temperature, and different carrier metals mixing are shown in Figures 2-6.

The pH variation to obtain $Be(OH)_2$ (Figure 2) shows that the best currents were obtained with a pH=9; at pH=7, the currents differ by 6.9% compared to pH=9, but these were more unstable. Whereas higher pH did not turn out to be optimal.

Figure 3 shows the calcination temperature variation. At 1000°C, the highest currents and beam stability are achieved; at higher temperatures, there is the best stability and produces 9.9% lower currents than 1000°C, but the smaller temperatures reduce 39.4% the currents compared to higher currents.

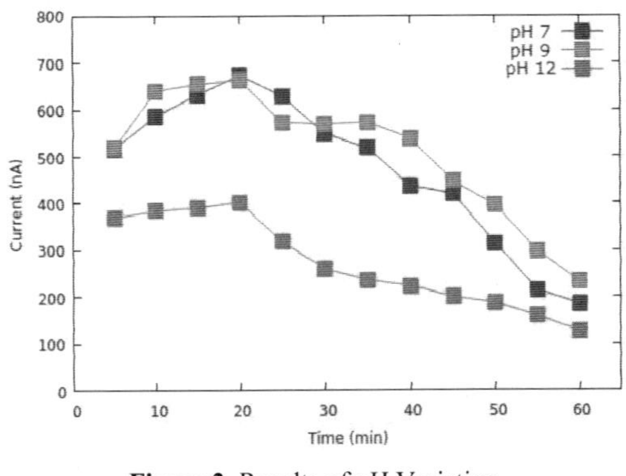

Figure 2. Results of pH Variation.

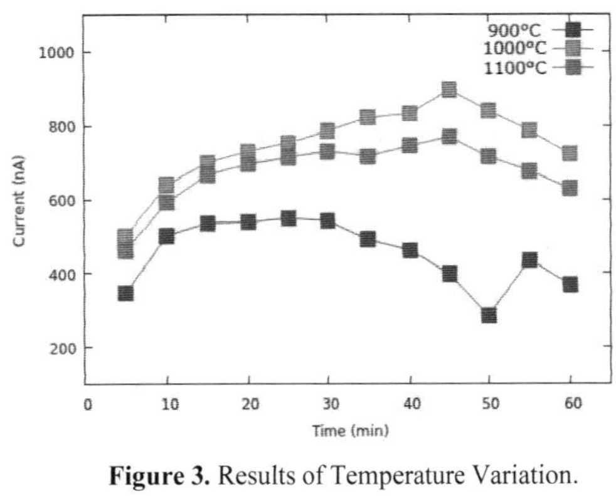

Figure 3. Results of Temperature Variation.

For different carrier metals (Figure 4), with iron (Fe), the lowest currents are obtained, and copper (Cu) increases them by 21.9%, both are unstable. On the contrary, niobium (Nb) and silver (Ag) generate the highest currents having an improvement of 53.8% and 57.6%, respectively, compared to Fe, and present good stabilities.

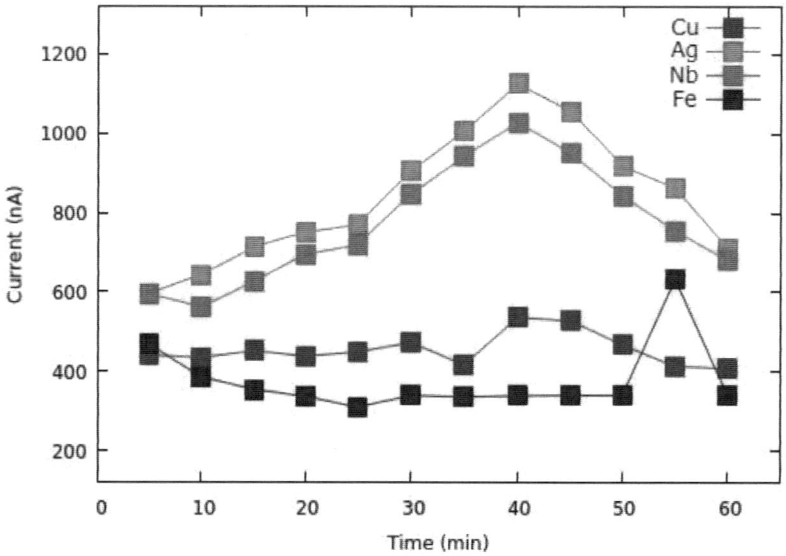

Figure 4. Results of Carrier Metal Variation.

By analyzing different proportions of the carrier metal used in the protocol and exchanging it for silver, the following results are obtained. For Nb (Figure 5), 1:1 and 1:2 ratios give lower currents; 1:3 ratio showed an increase in current of 92.2% compared to 1:1 ratio; higher ratios (1:4 and 1:5) duplicate the currents of 1:3 ratio; in the stability for ratios 1:4 and 1:5, no defined stability plateau is observed, however, the variations between currents are slight up to minute 40, which allows measuring ^{10}Be in natural samples. For Ag (Figure 6), with 1:3 ratio, the currents are small and have a difference of 4.7% with 1:4 ratio; but the best currents are achieved with 1:5 and 1:6 ratios, with a little difference between them (1.9%); all ratios present similar stability.

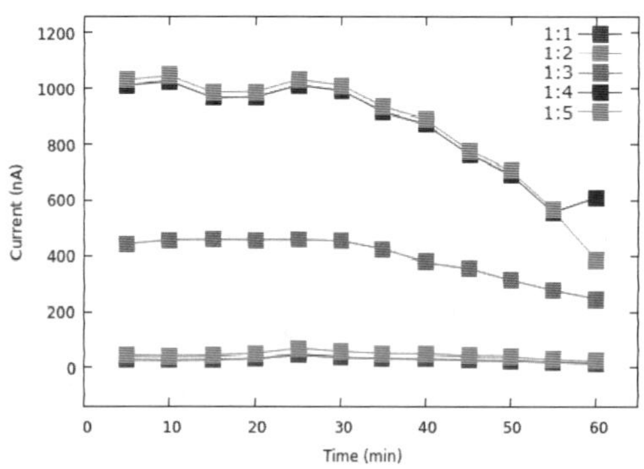

Figure 5. Results of the Niobium Proportions.

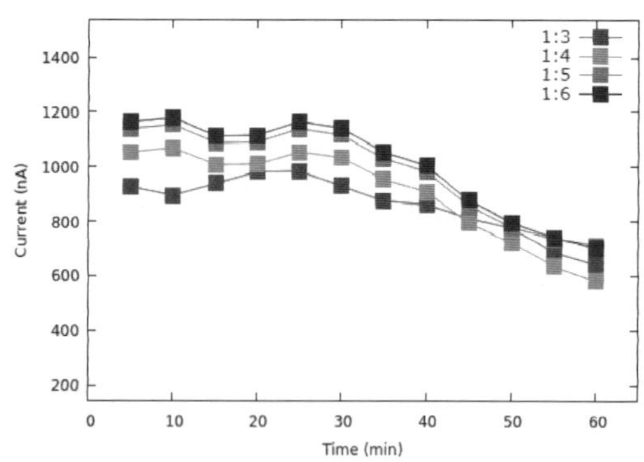

Figure 6. Results of the Silver Proportions.

Finally, the initial and optimized protocol was applied to real samples, the figure 7 the results with Ag show improvement in the currents and stability with a percentual difference of 25.6% compared to the initial protocol. The concentrations of ^{10}Be in the samples that were applied to the optimized protocol showed a lower uncertainty with respect to the initial protocol, 4 and 7%, respectively.

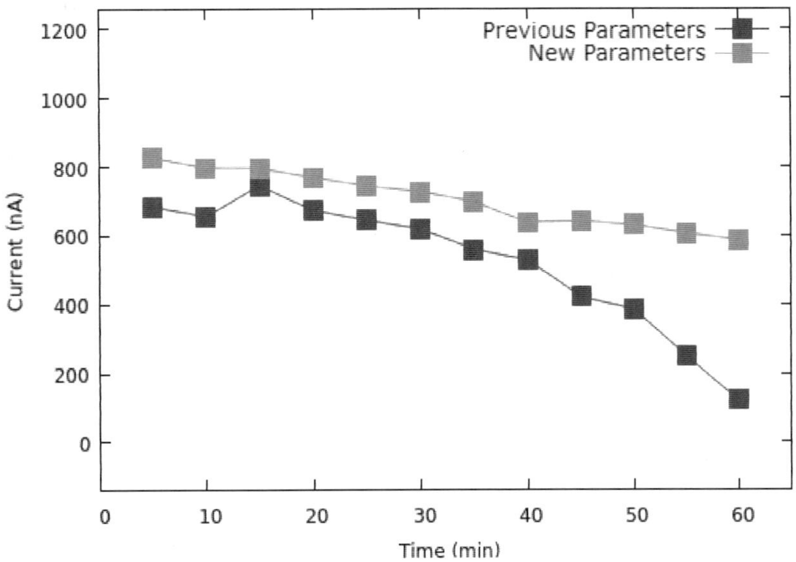

Figure 7. Comparison of Results with Nb and Ag.

4. Conclusion

The initial protocol (pH=9, calcination temperature at 1000°C and Nb mixture 1:4) showed great efficiency when obtaining high currents for ^{10}Be extraction, however, it was also found that changing from Nb to Ag carrier metal at 1:5 ratio achieved more significant beam currents and lower uncertainties in ^{10}Be concentration measurements. This change, when applied to natural samples, maintains the improvement observed in the blanks. For its use, a cost-benefit study with respect to Nb is recommended.

References

[1] S. Padilla, J.M. López-Gutiérrez, D.M.R. Sampath, T. Boski, J.M. Nieto, M. García-León, Determination of denudation rates by the measurement of meteoric 10Be in Guadiana river sediment samples (Spain) by low-energy AMS, J Environ Radioact. 189 (2018) 227–235. https://doi.org/10.1016/j.jenvrad.2018.04.016.

[2] G.M. Raisbeck, F. Yiou, Depositional rate and seasonal variation in 10Be precipitation, Nature. 282 (1979) 279–280.

[3] I. Graham, R. Ditchburn, B. Barry, Atmospheric deposition of 7Be and 10Be in New Zealand rain (1996-98), Geochim Cosmochim Acta. 67 (2003) 361–373. https://doi.org/10.1016/S0016-7037(02)01092-X.

[4] K.J. Kim, Y. Choi, Y.Y. Yoon, Monitoring 7Be and tritium in rainwater in Daejeon, Korea and its significance, Applied Radiation and Isotopes. 109 (2016) 470–473. https://doi.org/10.1016/j.apradiso.2015.11.094.

[5] C. Tiessen, D. Bemmerer, G. Rugel, R. Querfeld, A. Scharf, G. Steinhauser, S. Merchel, Accelerator mass spectrometry (AMS) for beryllium-7 measurements in smallest rainwater samples, J Radioanal Nucl Chem. 319 (2019) 965–973. https://doi.org/10.1007/s10967-018-6371-6.

[6] C.G. Méndez-García, S. Padilla, G. Rojas-López, R. García, C. Solís, E. Chávez, A. Huerta, K. De los Ríos, Atmospheric deposition of 10Be in Altzomoni rainfall, Nucl Instrum Methods Phys Res B. 536 (2023) 60–66. https://doi.org/10.1016/j.nimb.2022.12.020.

[7] L.R. McHargue, A.J.T. Jull, A. Cohen, Measurement of 10Be from Lake Malawi (Africa) drill core sediments and implications for geochronology, Palaeogeogr Palaeoclimatol Palaeoecol. 303 (2011) 110–119. https://doi.org/10.1016/j.palaeo.2010.02.012.

[8] M. Frank, B. Schwarz, S. Baumann, P.W. Kubik, M. Suter, A. Mangini, A 200 kyr record of cosmogenic radionuclide production rate and geomagnetic field intensity from 10Be in globally stacked deep-sea sediments, Earth Planet Sci Lett. 149 (1997) 121–129. https://doi.org/10.1016/s0012-821x(97)00070-8.

[9] B.S. Amin, D. Lal, B.L.K. Somayajulu, Chronology of marine sediments using the 10Be method: intercomparison with other methods, Geochim Cosmochim Acta. 39 (1975) 1187–1192. https://doi.org/10.1016/0016-7037(75)90060-5.

[10] J.A. Abreu, J. Beer, F. Steinhilber, M. Christl, P.W. Kubik, 10Be in ice cores and 14C in tree rings: Separation of production and climate effects, Space Sci Rev. 176 (2013) 343–349. https://doi.org/10.1007/s11214-011-9864-y.

[11] J. Jouzel, G. Raisbeck, J.P. Benoist, F. Yiou, C. Lorius, D. Raynaud, J.R. Petit, N.I. Barkov, Y.S. Korotkevitch, V.M. Kotlyakov, A comparison of deep Antarctic ice cores and their implications for climate between 65,000 and 15,000 years ago, Quat Res. 31 (1989) 135–150. https://doi.org/10.1016/0033-5894(89)90003-3.

[12] C.G. Méndez-García, G. Rojas-López, S. Padilla, C. Solís, E. Chávez, L. Acosta, A. Huerta, The impact of stable 27Al in 26Al/10Be meteoric ratio in PM2.5 from an urban area, J Environ Radioact. 246 (2022) 106832. https://doi.org/https://doi.org/10.1016/j.jenvrad.2022.106832.

[13] C.G. Méndez-García, S. Padilla, C. Solís, K. De los Ríos, E. Chávez, R. García, L. Acosta, A. Huerta, Meteoric 10Be concentrations in the center of Mexico, J Radioanal Nucl Chem. 322 (2019) 1455–1460. https://doi.org/10.1007/s10967-019-06841-x.

[14] S. Padilla Domínguez, Medidas de 10Be y 26Al en espectrometría de masas con acelerador de baja energía en el Centro Nacional de Aceleradores, (2015). http://hdl.handle.net/11441/32023 (accessed May 10, 2017).

[15] K. De Los Rios, C. Méndez-Garcia, S. Padilla, C. Solis, E. Chávez, A. Huerta, L. Acosta, Characterization of the LEMA isotope separator to measure concentrations of 10Be from atmospheric filters, J Phys Conf Ser. 1078 (2018). https://doi.org/10.1088/1742-6596/1078/1/012009.

[16] S. Padilla, J.M. López-Gutiérrez, G. Manjón, R. García-Tenorio, J.A. Galván, M. García-León, Meteoric 10Be in aerosol filters in the city of Seville, J Environ Radioact. 196 (2019) 15–21. https://doi.org/10.1016/j.jenvrad.2018.10.009.

[17] C. Solís, E. Chávez-Lomelí, M.E. Ortiz, A. Huerta, E. Andrade, E. Barrios, A new AMS facility in Mexico, Nucl Instrum Methods Phys Res B. 331 (2014) 233–237. https://doi.org/10.1016/j.nimb.2014.02.015.

[18] K. Nishiizumi, M. Imamura, M.W. Caffee, J.R. Southon, R.C. Finkel, J. McAninch, Absolute calibration of 10Be AMS standards, Nucl Instrum Methods Phys Res B. 258 (2007) 403–413. https://doi.org/10.1016/j.nimb.2007.01.297.

[19] E.C. Calvo, F.J. Santos, J.M. López-Gutiérrez, S. Padilla, M. García-León, J. Heinemeier, C. Schnabel, G. Scognamiglio, Status report of the 1 MV AMS facility at the Centro Nacional de Aceleradores, Nucl Instrum Methods Phys Res B. 361 (2015) 13–19. https://doi.org/10.1016/j.nimb.2015.02.022.

[20] G. Scognamiglio, Optimization of 10Be and 26Al detection with low-energy accelerator mass

spectrometry, University of Seville, 2017

Journal of Physics: Conference Series **2619** (2023) 012018 doi:10.1088/1742-6596/2619/1/012018

Description of the low-mass meson-like spectrum in non-perturbative QCD

Tochtli Yépez-Martínez[1], **Peter O. Hess**[2,3] **and Osvaldo Civitarese**[4]

[1] Instituto de Educación Media Superior de la Ciudad de México, Plantel General Lázaro Cárdenas del Río, Av. Jalalpa Norte 120, Colonia Jalalpa El Grande, C.P. 01377, Alcaldía Alvaro Obregón, Ciudad de México, México.

[2] Instituto de Ciencias Nucleares, Universidad Nacional Autónoma de México, Ciudad Universitaria, Circuito Exterior S/N, A.P. 70-543, 04510 M´exico D.F. Mexico.

[3] Frankfurt Institute for Advanced Studies, J. W. von Goethe University, Hessen, Germany.

[4] Departamento de Física, Universidad Nacional de La Plata, C.C. 67 (1900), La Plata, Argentina, and IFLP-CONICET.

E-mail: tochtlicuauhtli.yepez@iems.edu.mx, hess@nucleares.unam.mx, osvaldo.civitarese@fisica.unlp.edu.ar

Abstract. In this work we shall explore the identification and use of effective degrees of freedom, for the description of the non-perturbative regime of QCD. The starting Hamiltonian is the effective Coulomb plus linear potential. In order to build the spectrum of effective quark degrees of freedom, starting from arbitrary chosen quark masses, we shall pre-diagonalized the Hamiltonian. Then, we shall include quark-pair correlations and treat them by applying Bogoliubov transformations. The resulting quasiparticle excitations are then used to construct non-interacting two quasi-quark states with energies up to 2 GeV, which are compared to the observed meson-like states in the same energy domain.

1. Introduction

The Quantum Chromodynamics (QCD) is the theory of strong interactions [1, 2, 3], its low energy regime is non-perturbative and in consequence one is forced to apply certain approximations in order to describe hadronic states. Thus the efforts focus in the construction of effective, low-energy approximations to QCD and to find out methods to describe its non-perturbative domain. In this work we investigate the properties of a phenomenological model inspired by the canonical approach to QCD in the physical, Coulomb gauge quantization.

The underlying interactions in Coulomb gauge are dominated by the Instantaneous Color Coulomb Interaction (ICCI) acting between color charges. It has been shown in [4, 5, 6, 7] that the ICCI becomes confining, *i.e.*, proportional to the distance between the external color charges. Beyond the effective quark and antiquark spectrum, the correlations between quark-antiquark pairs lead to meson-like excitations which may be treated as a many-body problem.

In hadronic models, such approximations are driven by phenomenological considerations rather than by QCD itself. Examples of these are the bag model [8, 9], the algebraic models [10, 11, 12, 13] and the constituent-quark model [14]. For these models a number of experimental data points should be considered in order to fit their parameters.

The QCD Hamiltonian in the Coulomb gauge, is a good starting point for the application of many body techniques [15, 16]. The harmonic oscillator basis allows not only to obtain analytic expressions for matrix elements, as shown in [17, 18], but also helps to reduce the size of the matrix to be diagonalized. A cut-off in the harmonic oscillator basis is related to the maximal number of oscillator quanta included in the calculations. The use of the oscillator basis introduces a new way to describe hadrons as a combination of discrete states. This new approach requires another way of performing the renormalization of running constituent quark masses and of the interaction constant. By looking at correlations between pairs of quarks which may result in a superfluid low energy regime, the observed spectrum of hadrons could be interpreted in terms of quasiparticle excitations. This is the purpose of the present work.

In Section 2, we proceed gradually by treating the QCD Hamiltonian in the Coulomb gauge in order to construct the effective quark spectrum. In Section 3, we apply the BCS transformations and solve the corresponding flavor dependent equations, in order to get the relevant parameters of the model, that is the up/down and strange quasi-quark energies and gaps. In Section 4, we present the dependence of the solutions in terms of the dimension of the basis. The conclusions are drawn in Section 5.

2. Quarks and antiquarks as effective degrees of freedom.

The first step for the implementation of the BCS method consists in the definition of the single particle field. In the case of QCD the literature is rich in the description of such fields [19, 20, 21, 22, 23, 24, 25, 17].

2.1. The QCD Hamiltonian in the Coulomb gauge.

We start from the QCD Hamiltonian in its canonical Coulomb gauge representation [3, 2],

$$\begin{aligned}\boldsymbol{H}^{QCD} &= \int \left\{ \frac{1}{2}\left[\mathcal{J}^{-1}\boldsymbol{\Pi}^{tr}\cdot\mathcal{J}\boldsymbol{\Pi}^{tr} + \mathcal{B}\cdot\mathcal{B}\right] - \overline{\psi}\left(-i\gamma\cdot\nabla + m\right)\psi - g\overline{\psi}\gamma\cdot A\psi \right\} d\mathbf{x} \\ &+ \frac{1}{2}g^2 \int \mathcal{J}^{-1}\rho^c(\mathbf{x})\langle c,\mathbf{x}|\frac{1}{\nabla\cdot\mathcal{D}}(-\nabla^2)\frac{1}{\nabla\cdot\mathcal{D}}|c'\mathbf{y}\rangle\mathcal{J}\rho^{c'}(\mathbf{y})d\mathbf{x}d\mathbf{y}\ , \qquad (1)\end{aligned}$$

which has been widely studied in the past for the description of several properties of QCD at low energy [4, 5, 6, 7, 26, 27, 28, 29, 30, 31, 32]. The last term in the Hamiltonian of Eq. (1) includes the relevant interactions between quarks and gluons *i.e.*, the *QCD Instantaneous color-Coulomb Interaction* (QCD-IcCI). At low energy the effects of dynamical gluons in the QCD-IcCI can be represented by the interaction $V(|\mathbf{x}-\mathbf{y}|) = -\frac{V_C}{|\mathbf{x}-\mathbf{y}|} + V_L|\mathbf{x}-\mathbf{y}|$, which is obtained from a self-consistent treatment of the interaction between color charge-densities [5, 6].

Here, we have analysed the quark sector of the Hamiltonian of Eq. (1) and taken the effective confining interaction $V(|\mathbf{x}-\mathbf{y}|)$ to describe the low energy interaction between color charge densities. We write

$$\begin{aligned}\boldsymbol{H}^{QCD}_{eff} &= \int \left\{\boldsymbol{\psi}^\dagger(\mathbf{x})(-i\boldsymbol{\alpha}\cdot\boldsymbol{\nabla} + \beta m)\boldsymbol{\psi}(\mathbf{x})\right\} d\mathbf{x} - \frac{1}{2}\int \rho_c(\mathbf{x})V(|\mathbf{x}-\mathbf{y}|)\rho^c(\mathbf{y})d\mathbf{x}d\mathbf{y} \\ &= \boldsymbol{K} + \boldsymbol{H}_{\mathrm{Coul}}\ , \qquad (2)\end{aligned}$$

where $\rho^c(\mathbf{x}) = \psi^\dagger(\mathbf{x})T^c\psi(\mathbf{x})$ is the quark and antiquark charge-density and $T^c(T_c)$ are the SU(3) color generators, because of the non-abelian structure of QCD. The first term in Eq. (2) is the kinetic energy, and the second term is the QCD-IcCI in its simplified form. The fermion field $\psi^\dagger(\mathbf{x})$, whose quantization is expanded in terms of creation and annihilation operators in the harmonic oscillator basis

$$\psi^\dagger(\mathbf{x}) \;=\; \sum_{Nlm_l,\sigma CF} R^*_{Nl}(x)Y^*_{lm_l}(\hat{\mathbf{x}})\chi^\dagger_\sigma\left(\boldsymbol{q}^\dagger_{\frac{1}{2},Nlm_l,\sigma CF}+\boldsymbol{q}^\dagger_{-\frac{1}{2},Nlm_l,\sigma CF}\right)\,. \tag{3}$$

with $x=|\mathbf{x}|$ and $R_{Nl}(x)=N_{Nl}\exp(-\frac{B_0x^2}{2})x^lL^{l+\frac{1}{2}}_{\frac{N-l}{2}}(B_0x^2)$, where L^λ_n is an associated Laguerre polynomial and $(\sqrt{B_0})^{-1}$ is the oscillator length. The index τ denotes upper ($\tau=\frac{1}{2}$) and lower ($\tau=-\frac{1}{2}$) components of the Dirac spinors in the Dirac-Pauli representation of the Dirac matrices, and σ, C, F denote spin, color and flavor intrinsic degrees of freedom, respectively.

The use of the harmonic oscillator basis requires the diagonalization of the kinetic term (2), which is performed by using the total spin $J=l\pm\frac{1}{2}$, for a given maximal number of quanta $N=N_{\rm cut}$. For this, we introduce a general transformation to a basis of effective operators,

$$\boldsymbol{q}^\dagger_{\tau(Nl)JM_JCF} \;=\; \sum_{\lambda\pi k}\left(\alpha^{J,T}_{\tau(Nl),\lambda\pi k}\right)^*\boldsymbol{Q}^\dagger_{\lambda\pi kJM_JCF}\;\delta_{\pi,(-1)^{\frac{1}{2}-\tau+l}}\,. \tag{4}$$

where the value $\lambda=+\frac{1}{2}$ will refer to effective quarks and the value $\lambda=-\frac{1}{2}$ to effective antiquarks, *i.e.* $\boldsymbol{Q}^\dagger_{\frac{1}{2}\pi kJM_JCF}\to\boldsymbol{b}^\dagger_{\pi kJC(Y,T),M_JM_CM_T}$ and $\boldsymbol{Q}^\dagger_{-\frac{1}{2}\pi kJM_JCF}\to\boldsymbol{d}_{\pi kJC(Y,T),M_JM_CM_T}$.

With the use of the harmonic oscillator basis and the pre.-diagonalization of the kinetic term, we have identified effective quarks and antiquarks as a linear combinations of the bare quarks and antiquarks. In terms of these effective quarks and antiquarks, the kinetic energy term acquires the following structure

$$K=\sum_{k\pi\gamma}\varepsilon_{k\pi\gamma}\sum_\mu\left(\boldsymbol{b}^\dagger_{k\pi\gamma\mu}\boldsymbol{b}^{k\pi\gamma\mu}-\boldsymbol{d}^{k\pi\gamma\mu}\boldsymbol{d}^\dagger_{k\pi\gamma\mu}\right)\,, \tag{5}$$

The indices indicate the principal number ($k=1,2,\dots$) which run over all orbital states and the parity ($\pi=\pm$) while γ, μ are short hand notation for the particle spin, color and flavor hypercharge and isospin representations $\gamma=\{J,C,(Y,T)\}$ and their magnetic projections $\mu=\{M_J,M_C,M_T\}$, respectively.

The QCD-IcCI term, in its simplified form ($\boldsymbol{H}_{\rm Coul}$), rewritten in terms of effective quarks and antiquarks operators is given by [18]

$$\begin{aligned}\boldsymbol{H}_{Coul} =\; & -\frac{1}{2}\sum_L\sum_{\lambda_i\mathbf{q}_i}V^L_{\{\lambda_i\mathbf{q}_i\}}\Big(\left[\mathcal{F}_{\lambda_1\mathbf{q}_1,\lambda_2\mathbf{q}_2;\gamma_{f_0}}\mathcal{F}_{\lambda_3\mathbf{q}_3,\lambda_4\mathbf{q}_4;\bar\gamma_{f_0}}\right]^{\gamma_0}_{\mu_0}+\left[\mathcal{F}_{\lambda_1\mathbf{q}_1,\lambda_2\mathbf{q}_2;\gamma_{f_0}}\mathcal{G}_{\lambda_3\mathbf{q}_3,\lambda_4\mathbf{q}_4;\bar\gamma_{f_0}}\right]^{\gamma_0}_{\mu_0}\\ & +\left[\mathcal{G}_{\lambda_1\mathbf{q}_1,\lambda_2\mathbf{q}_2;\gamma_{f_0}}\mathcal{F}_{\lambda_3\mathbf{q}_3,\lambda_4\mathbf{q}_4;\bar\gamma_{f_0}}\right]^{\gamma_0}_{\mu_0}+\left[\mathcal{G}_{\lambda_1\mathbf{q}_1,\lambda_2\mathbf{q}_2;\gamma_{f_0}}\mathcal{G}_{\lambda_3\mathbf{q}_3,\lambda_4\mathbf{q}_4;\bar\gamma_{f_0}}\right]^{\gamma_0}_{\mu_0}\Big)\end{aligned} \tag{6}$$

where we have compacted the single particle orbital number, parity and irreps into the short-hand notation $\mathbf{q}_i=k_i\pi_{q_i}\gamma_{q_i}$, and use for the (flavor-less) quantum numbers of the intermediate coupling in the interaction $\gamma_{f_0}=\{L(11)(0,0)\}$ and for their magnetic projections $\mu_{f_0}=\{M_L,M_C,0\}$. The conjugate representations satisfy $\bar\gamma_{f_0}=\gamma_{f_0}$ and $\bar\mu_{f_0}=\{-M_L,\bar M_C,0\}$. For the total couplings (upper index) and magnetic numbers (lower index) of the interaction, we have used $\gamma_0=\{0,(00),(0,0)\}$ and $\mu_0=\{0,0,0\}$ respectively. The operators $\mathcal{F}$ and $\mathcal{G}$ are given by

$$\begin{aligned}\mathcal{F}_{\lambda_1\mathbf{q}_1,\lambda_2\mathbf{q}_2;\gamma_{f_0},\mu_{f_0}} &= \frac{1}{\sqrt{2}}\left\{\delta_{\lambda_1,\frac{1}{2}}\delta_{\lambda_2,\frac{1}{2}}\left[\boldsymbol{b}^\dagger_{\mathbf{q}_1}\otimes\boldsymbol{b}_{\bar{\mathbf{q}}_2}\right]^{\gamma_{f_0}}_{\mu_{f_0}}-\delta_{\lambda_1,-\frac{1}{2}}\delta_{\lambda_2,-\frac{1}{2}}\left[\boldsymbol{d}_{\bar{\mathbf{q}}_1}\otimes\boldsymbol{d}^\dagger_{\bar{\mathbf{q}}_2}\right]^{\gamma_{f_0}}_{\mu_{f_0}}\right\}\\ \mathcal{G}_{\lambda_1\mathbf{q}_1,\lambda_2\mathbf{q}_2;\gamma_{f_0},\mu_{f_0}} &= \frac{1}{\sqrt{2}}\left\{\delta_{\lambda_1,-\frac{1}{2}}\delta_{\lambda_2,\frac{1}{2}}\left[\boldsymbol{d}_{\mathbf{q}_1}\otimes\boldsymbol{b}_{\bar{\mathbf{q}}_2}\right]^{\gamma_{f_0}}_{\mu_{f_0}}-\delta_{\lambda_1,\frac{1}{2}}\delta_{\lambda_2,-\frac{1}{2}}\left[\boldsymbol{b}^\dagger_{\mathbf{q}_1}\otimes\boldsymbol{d}^\dagger_{\bar{\mathbf{q}}_2}\right]^{\gamma_{f_0}}_{\mu_{f_0}}\right\}\,.\end{aligned} \tag{7}$$

Journal of Physics: Conference Series **2619** (2023) 012018 doi:10.1088/1742-6596/2619/1/012018

In this basis, and using the above introduced states, the matrix elements of the interaction are given by

$$\begin{aligned} V^L_{\{\lambda_i\pi_i k_i J_i Y_i T_i\}} &= \sum_{\tau_i N_i l_i} V^L_{\{N_i l_i J_i\}}\, \alpha^{J_1,T_1}_{\tau_1(N_1 l_1),\lambda_1,\pi_1,k_1} \alpha^{J_2,T_2}_{\tau_2(N_2 l_2),\lambda_2,\pi_2,k_2} \alpha^{J_3,T_3}_{\tau_3(N_3 l_3),\lambda_3,\pi_3,k_3} \alpha^{J_4,T_4}_{\tau_4(N_4 l_4),\lambda_4,\pi_4,k_4} \\ &\times \delta_{\tau_1\tau_2}\delta_{\tau_3\tau_4}\, \delta_{\pi_1,(-1)^{\frac{1}{2}-\tau_1+l_1}} \delta_{\pi_2,(-1)^{\frac{1}{2}-\tau_2+l_2}} \delta_{\pi_3,(-1)^{\frac{1}{2}-\tau_3+l_3}} \delta_{\pi_4,(-1)^{\frac{1}{2}-\tau_4+l_4}} \\ &\times (-1)^{\frac{1}{3}+\frac{Y_1}{2}+T_1} \frac{\sqrt{2T_1+1}}{\sqrt{3}} \delta_{T_2T_1}\delta_{Y_2Y_1}\, (-1)^{\frac{1}{3}+\frac{Y_3}{2}+T_3} \frac{\sqrt{2T_3+1}}{\sqrt{3}} \delta_{T_4T_3}\delta_{Y_4Y_3}. \end{aligned} \tag{8}$$

The matrix elements in the harmonic oscillator basis ($V^L_{\{N_i l_i J_i\}}$) are analytic and actually easy to compute.

3. Bogoliubov Transformation and the BCS method.

Here, starting form the effective quark and antiquarks basis, we apply the *Bogoliubov Transformation* [15, 16], in order to approximately diagonalize part of the effective QCD interaction, Eq. (2).

Herewith we shall apply the transformations for each quark-flavor, separately. Thus, the reference (ground state) state will be a quark condensate with a definite flavor. Then, the Hamiltonian of Eq.(2) is written in terms of quasi-quark operators and the standard conditions of the BCS theory are applied to it by asking the one-quasiparticle term (H_{11}) to be diagonal, the pairs terms ($H_{20}+H_{02}$) to vanish. A crucial step in the treatment, leading to the transformation between ordinary particles (in this case fermions like the quarks and antiquarks) to quasiparticles is the replacement of the ordinary vacuum $|0\rangle$ by the BCS vacuum $|BCS\rangle$,

$$|BCS\rangle \sim |0\rangle + \sum_{k_1\pi_1\gamma_1} z_{k_1\pi_1\gamma_1} \left[\boldsymbol{b}^\dagger_{k_1\pi_1\gamma_1} \otimes \boldsymbol{d}^\dagger_{k_1\pi_1\bar{\gamma}_1} \right]^{\gamma_0}_{\mu_0} |0\rangle \tag{9}$$

The quasiparticle excitation $\boldsymbol{B}^\dagger_{k_i\pi_i,\gamma_i\mu_i}|BCS\rangle$ operator has the quantum numbers of the quark, and the quasiparticle excitation $\boldsymbol{D}^\dagger_{k_i\pi_i,\gamma_i\mu_i}|BCS\rangle$ operator has the quantum numbers of the antiquark, being $\boldsymbol{B}^\dagger_{k_i\pi_i,\gamma_i\mu_i}$ and $\boldsymbol{D}^\dagger_{k_i\pi_i,\gamma_i\mu_i}$ the creation operators for quasi-quarks and quasi-antiquarks, respectively.

At the same time the following conditions:

$$\boldsymbol{B}^{k_i\pi_i,\gamma_i\mu_i}|BCS\rangle = \boldsymbol{D}_{k_i\pi_i,\gamma_i\mu_i}|BCS\rangle = 0, \tag{10}$$

$$\left\{ \boldsymbol{B}^{k_i\pi_i,\gamma_i\mu_i}, \boldsymbol{B}^\dagger_{k_j\pi_j,\gamma_j\mu_j} \right\} = \left\{ \boldsymbol{D}_{k_i\pi_i,\gamma_i\mu_i}, \boldsymbol{D}^{\dagger k_j\pi_j,\gamma_j\mu_j} \right\} = \delta_{ij} \tag{11}$$

and

$$\left\{ \boldsymbol{B}^{k_i\pi_i,\gamma_i\mu_i}, \boldsymbol{D}^{\dagger k_j\pi_j,\gamma_j\mu_j} \right\} = \left\{ \boldsymbol{B}^{k_i\pi_i,\gamma_i\mu_i}, \boldsymbol{D}_{k_i\pi_i,\gamma_i\mu_i} \right\} = 0 \tag{12}$$

should be obeyed.

The next step in our derivation consists of taking normal order respect to the quasiparticle vacuum and collecting the different contributions to the Hamiltonian with constant terms H_{00}, one creation-one annihilations terms H_{11}, two-quasiparticle terms H_{20} and H_{02}.

The value of the gap is extracted, for each channel, by solving the set of BCS equations (see Section 3.1) and the remanent of the transformed Hamiltonian may be treated in the RPA basis to describe correlations between pairs of quasiparticles [33].

3.1. BCS equations.
The terms H_{11} and $H_{20}+H_{02}$ can be ordered in terms of the following variables

$$\begin{aligned} X_{k_1} &= u_{k_1}^2 - v_{k_1}^2 \\ Y_{k_1} &= 2u_{k_1}v_{k_1} \end{aligned} \tag{13}$$

which depend only on the quasiparticle operator indices, k_i. Here, we are using a short-hand notation $k_i = k_i\pi_i, \gamma_i$. The interaction terms are also ordered in terms of the structures $(u_{k_2}^2 - v_{k_2}^2)$ and $(u_{k_2}v_{k_2})$, being k_2 an internal index. The parameters u_k and v_k are determined self-consistently.

The resulting equations are the well known BCS-equations

$$\Sigma_{k_1}X_{k_1} + \Delta_{k_1}Y_{k_1} = E_{k_1} \tag{14}$$

$$-\Delta_{k_1}X_{k_1} + \Sigma_{k_1}Y_{k_1} = 0 \tag{15}$$

where Σ_{k_1}, Δ_{k_1} and the quasiparticle energy E_{k_1} are given by

$$\Sigma_{k_1} = \varepsilon_{k_1} + \bar{V}^{\Sigma}_{k_1k_2k_2k_1}(u_{k_2}^2 - v_{k_2}^2) \tag{16}$$

$$\begin{aligned} \Delta_{k_1} &= \bar{V}^{\Delta}_{k_1k_2k_2k_1}(u_{k_2}v_{k_2}) \\ E_{k_1} &= \sqrt{\Sigma_{k_1}^2 + \Delta_{k_1}^2} \end{aligned} \tag{17}$$

with

$$\begin{aligned} &\bar{V}^{\Sigma}_{k_1\pi_1J_1Y_1T_1,\ k_2\pi_2J_2Y_2T_2,\ k_2\pi_2J_2Y_2T_2,\ k_1\pi_1J_1Y_1T_1} = \\ &-\frac{1}{2}\sum_L\sum_{\lambda_i}\left(\frac{1}{2}\right)\frac{\sqrt{8(2L+1)}}{9}\frac{(-1)^{L+J_2-J_1}}{2J_1+1} \\ &\times\Bigg\{\sum_{\tau_iN_il_i} V^L_{\{N_il_iJ_i\}}\alpha^{J_1,T_1}_{\tau_1(N_1l_1),\lambda_1,\pi_1,k_1}\alpha^{J_2,T_2}_{\tau_2(N_2l_2),\lambda_2,\pi_2,k_2}\alpha^{J_3,T_3}_{\tau_3(N_3l_3),\lambda_3,\pi_3,k_3}\alpha^{J_4,T_4}_{\tau_4(N_4l_4),\lambda_4,\pi_4,k_4} \\ &\times\delta_{\tau_1\tau_2}\delta_{\tau_3\tau_4}\delta_{\pi_1,(-1)^{\frac{1}{2}-\tau_1+l_1}}\delta_{\pi_2,(-1)^{\frac{1}{2}-\tau_2+l_2}}\delta_{\pi_3,(-1)^{\frac{1}{2}-\tau_3+l_3}}\delta_{\pi_4,(-1)^{\frac{1}{2}-\tau_4+l_4}}\Bigg\} \\ &\times(\delta_{\pi_1\pi_4})(\delta_{k_2k_3}\delta_{\pi_2\pi_3})(\delta_{J_2J_3}\delta_{J_1J_4})(\delta_{T_1T_2}\delta_{T_2T_3}\delta_{T_3T_4})(\delta_{Y_1Y_2}\delta_{Y_2Y_3}\delta_{Y_3Y_4}) \\ &\times(\delta_{\lambda_1+\frac{1}{2}}\delta_{\lambda_2+\frac{1}{2}}\delta_{\lambda_3+\frac{1}{2}}\delta_{\lambda_4+\frac{1}{2}} - \delta_{\lambda_1-\frac{1}{2}}\delta_{\lambda_2+\frac{1}{2}}\delta_{\lambda_3+\frac{1}{2}}\delta_{\lambda_4-\frac{1}{2}})\text{average} \end{aligned} \tag{18}$$

and

$$\begin{aligned} &\bar{V}^{\Delta}_{k_1\pi_1J_1Y_1T_1,\ k_2\pi_2J_2Y_2T_2,\ k_2\pi_2J_2Y_2T_2,\ k_1\pi_1J_1Y_1T_1} = \\ &-\frac{1}{2}\sum_L\sum_{\lambda_i}\left(\frac{1}{2}\right)\frac{\sqrt{8(2L+1)}}{9}\frac{(-1)^{L+J_2-J_1}}{2J_1+1} \\ &\times\Bigg\{\sum_{\tau_iN_il_i} V^L_{\{N_il_iJ_i\}}\alpha^{J_1,T_1}_{\tau_1(N_1l_1),\lambda_1,\pi_1,k_1}\alpha^{J_2,T_2}_{\tau_2(N_2l_2),\lambda_2,\pi_2,k_2}\alpha^{J_3,T_3}_{\tau_3(N_3l_3),\lambda_3,\pi_3,k_3}\alpha^{J_4,T_4}_{\tau_4(N_4l_4),\lambda_4,\pi_4,k_4} \\ &\times\delta_{\tau_1\tau_2}\delta_{\tau_3\tau_4}\delta_{\pi_1,(-1)^{\frac{1}{2}-\tau_1+l_1}}\delta_{\pi_2,(-1)^{\frac{1}{2}-\tau_2+l_2}}\delta_{\pi_3,(-1)^{\frac{1}{2}-\tau_3+l_3}}\delta_{\pi_4,(-1)^{\frac{1}{2}-\tau_4+l_4}}\Bigg\} \\ &\times(\delta_{\pi_1\pi_4})(\delta_{k_2k_3}\delta_{\pi_2\pi_3})(\delta_{J_2J_3}\delta_{J_1J_4})(\delta_{T_1T_2}\delta_{T_2T_3}\delta_{T_3T_4})(\delta_{Y_1Y_2}\delta_{Y_2Y_3}\delta_{Y_3Y_4}) \\ &\times 2(\delta_{\lambda_1+\frac{1}{2}}\delta_{\lambda_2+\frac{1}{2}}\delta_{\lambda_3-\frac{1}{2}}\delta_{\lambda_4-\frac{1}{2}} + \delta_{\lambda_1-\frac{1}{2}}\delta_{\lambda_2+\frac{1}{2}}\delta_{\lambda_3-\frac{1}{2}}\delta_{\lambda_4+\frac{1}{2}})\ , \end{aligned} \tag{19}$$

Journal of Physics: Conference Series **2619** (2023) 012018 doi:10.1088/1742-6596/2619/1/012018

where the summation on the internal indices is performed.

These non-linear equations are then solved for each of the quark-flavors and they are known as state depended BCS-equations [15, 16], because Σ_{k_1} and Δ_{k_1} depend on the flavor.

The procedure to obtain the solutions of Eqs. (14) and (15) consist of the variation of the quantities u_{k_i} and v_{k_i}, such that the iteration stop when the correlation energy E_{k_1} reaches stability.

4. Results and Discussions.

In this section we shall present the results of our calculations. We have considered the effects associated to the renormalization of the parameters, *i.e.*, the quarks masses and the Coulomb and linear interactions. At the same time we have studied the effects of a correlated vacuum and the presence of a gap for each quark flavor by comparing the spectrum of the pre-diagonalization energies with the quasiparticle energies, as a function of the cut-off parameter (N_{cut}). At the end of this section, we have constructed the uncorrelated two-quasiparticle spectrum and compared it to the experimental ones [34]. The main aspect of the comparison between calculations and data will focus on the density of states, for which we shall investigate if the space of uncorrelated two-quasiparticle states is dense enough to establish connections with data.

4.1. Renormalization of the parameters.

The renormalization implemented takes into account a cut-off in the number of oscillator quanta N_{cut}, such a truncation is similar to a momentum cut-off regularization, but in a discrete basis. The aim of the renormalization procedure is to keep at least the lowest eigenvalue of the BCS equations E_{k_i} unchanged, for each quark flavor. Thus, we present the renormalization results which, in fact, does succeed in keeping the low-energy meson-like spectrum approximately cut-off-independent.

We shall proceed by studying the dependence of the parameters which enter into the definition of the interaction, which are V_C and V_L . Fig. 1 shows the dependence of the parameters V_C and V_L upon the value of the cutoff, N_{cut}, of the radial basis. In doing so we have varied both V_C and V_L such that the resulting values of the gaps remain constant.

In Fig. 1, it is seen that the strength of the renormalized Coulomb interaction increases as a function of N_{cut}. This change is faster than the decrease observed in the strength of the renormalized linear interaction.

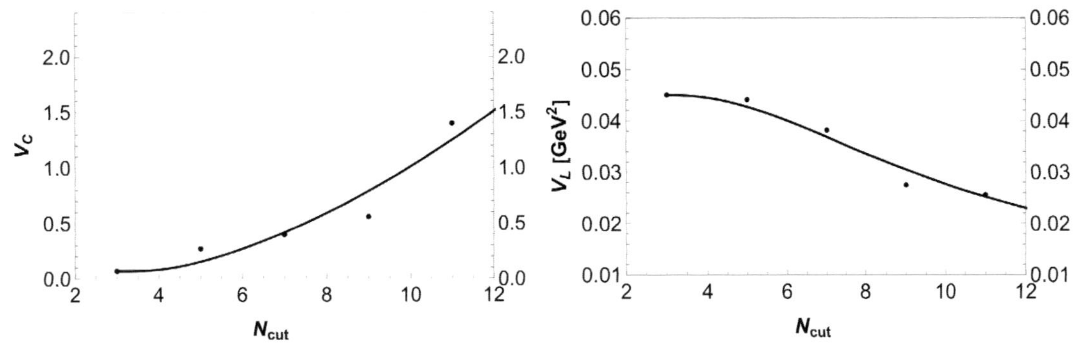

Figure 1: Renormalization of the Coulomb V_C (left) and linear interaction interaction V_L (right): dependence of the parameters upon the size of the radial basis (N_{cut}). The actual values are represented by dots, the line is to guide the eye. We are using natural units trough the text.

For the renormalization of the up/down quark mass m_q, we have used a starting value of $m_q = 0.050GeV$. In the case of the strange quark mass, we have used the highest value for which a gap in the strange sector is not zero, this is at about $m_s = 0.140GeV$, for higher values than this, there is no gap in the strange quark sector.

4.2. Quasiparticle spectrum.

Proceeding in the same manner, with the couplings of the previous subsection, we have diagonalized the one quasiparticle sector of the Hamiltonian and solved the BCS equations (14) and (15). The results for the up/down quarks $T = \frac{1}{2}$ and strange quark $T = 0$ are shown in Figure 2.

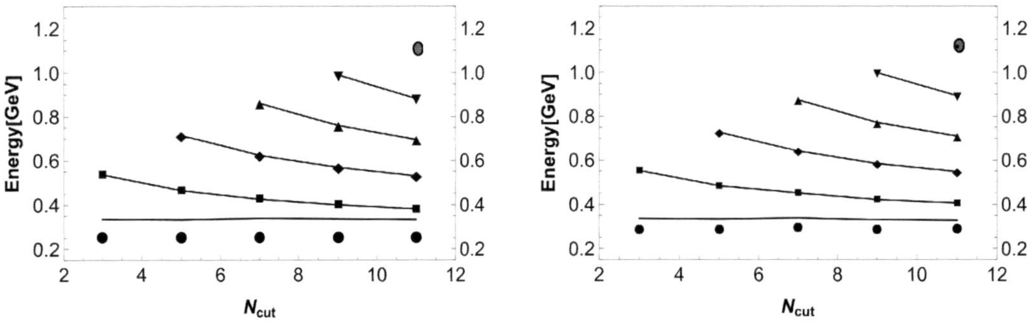

Figure 2: Prediagonalization energies (symbols) for $T = \frac{1}{2}$ (left) and $T = 0$ (right) states and Quasiparticle (solid lines) energies versus N_{cut}.

From the comparison between the energies obtained by diagonalizing the one particle sector of the Hamiltonian and those corresponding to the quasiparticles, we may conclude that in both cases the spectrum reaches a harmonic limit for large values of N_{cut}. The constancy of the gap for the up/down and strange quarks is well illustrated by the results shown in Figure 2, where the gap in the case of $T = \frac{1}{2}$ states is $\Delta \approx 0.2$ GeV and for $T = 0$ its value is $\Delta \approx 0.15$ GeV .

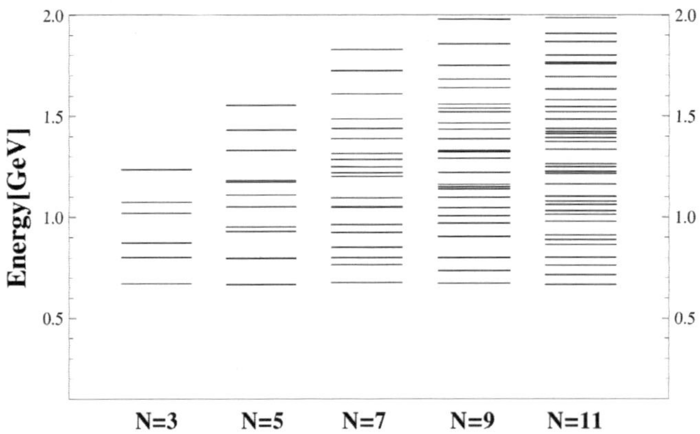

Figure 3: Two-Quasiparticle meson-like spectrum, for pairs of up and down quasiquarks, isospin $T = 0, 1$ states, vs N_{cut}.

4.2.1. Two quasiparticle spectrum as a function of N_{cut} Meson states, of positive and negative parities, are described as two-quasiparticle states. In Figure 3 the spectrum of two-quasiparticles, for the subspace $T = 0, 1$, is shown as a function of N_{cut}. The density of states increases as the

Journal of Physics: Conference Series **2619** (2023) 012018 doi:10.1088/1742-6596/2619/1/012018

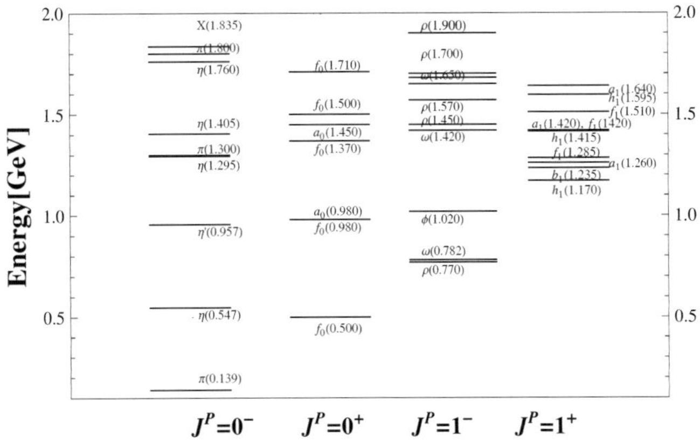

Figure 4: Experimental meson spectrum for isospin $T = 0, 1$, up to 2 Gev.

number of states in the basis increases. The density observed for $N_{cut} = 9, 11$ seems to be similar to that observed in the experimental spectrum shown in Figure 4. The figure is not meant to be compared with data but rather show the gross features of the spectrum. Considering that the theoretical results have been obtained at the quasiparticle level, that is without including residual interactions between pairs of quasiparticles, the overall tendency of them follows that of the experiment. This is particularly true for the sector of medium and high energies. This is encouraging because the addition of the residual terms of the interaction between pairs of quasiparticles, when treated in the context of non-perturbative linearization method, like the RPA approach, could certainly improve the agreement.

5. Conclusions.

In this work we have treated the Coulomb plus linear QCD Hamiltonian by applying non-perturbative techniques like the BCS method. A renormalization of the mass and interaction parameters was performed. The stability of the results was tested by increasing the dimension of the radial basis used in the calculations. We have calculated the gaps and quasiparticle energies for up/down and strange sectors. The two-quasiparticle spectrum for $T = \frac{1}{2}$ and $T = 0$, that is mesons and kaons like states, show features similar to those exhibited by the experiments.

It is expected that by including interactions between pairs of quasiparticles, would allow for a more detailed correspondence between theoretical and experimental results. Work is in progress about the use of the RPA methods in this context.

Acknowledgments

P.O.H. acknowledges financial support from PAPIIT-DGAPA (IN100421). O.C acknowledges the support of the CONICET and of the ANPCyT of Argentina (PIP-616).

[1] S. Weinberg, *The Quantum Theory of Fields* (Vol. II, Cambridge University Press, 1996).
[2] T. D. Lee, *Particle Physics and Introduction to Field Theory* (Harwood Academic Publishers, New York, 1981).
[3] N. H. Christ and T. D. Lee, Phys. Rev. D **22**, 939 (1980).
[4] A. Szczepaniak, E. S. Swanson, C. R. Ji and S. R. Cotanch, Phys. Rev. Lett. **76**, 2011 (1996).
[5] A. P. Szczepaniak and E. Swanson, Phys. Rev. D **65**, 025012 (2001).
[6] C. Feuchter and H. Reinhardt, Phys. Rev. D **70**, 105021 (2004).
[7] H. Reinhardt, D. R. Campagnari and A. P. Szczepaniak, Phys. Rev. D **84**, 045006 (2011).
[8] T. DeGrand, R. L. Jaffe, K. Johnson, and J. Kiskis, Phys. Rev. D **12**, 2060 (1975).
[9] W. Greiner, S. Schramm, and E. Stein, Quantum Chromodynamcis (Spiriner, Heidelberg, 2002).
[10] R. Bijker, F. Iachello, and A. Leviatan, Ann. Phys. (NY) **236**, 69 (1994).

[11] T. Yepez-Martinez, O. Civitarese and P. O. Hess, Int. J. Mod. Phys. E **25**, 1650067 (2016).
[12] T. Yepez-Martinez, O. Civitarese and P. O. Hess, Int. J. Mod. Phys. E **26**, 1750012 (2017).
[13] U. I. Ramirez-Soto, O. A. Rico-Trejo, T. Yépez-Martínez, P. O. Hess, A. Weber and O. Civitarese, J. Phys. G: Nucl. Part. Phys. **48**, 085013 (2021).
[14] N. Isgur and G. Karl, Phys. Rev. D **18**, 4187 (1978).
[15] P. Ring and P. Schuck, *The Nuclear Many Body Problem* (Springer, Heidelberg, 1980).
[16] A. L. Fetter and J. D. Walecka, *Quantum Theory of Many-Particle Systems* (Dover, New York, 2003).
[17] T. Yépez-Martínez, P. O. Hess, A. P. Szczepaniak and O. Civitarese, Phys. Rev. C **81**, 045204 (2010).
[18] D. A. Amor-Quiroz, T. Yépez-Martínez, P. O. Hess, O. Civitarese, and A. Weber, Int. J. Mod. Phys. E **26**, 1750082 (2017)
[19] J. R. Finger and J. E. Mandula, Nucl. Phys. B**199**, 168 (1982)
[20] S. L. Adler and A. C. Davis, Nucl. Phys. B**244**, 469 (1984)
[21] A. Le Yaouanc, L. Oliver, O. Pène, and J.-C. Raynal, Phys. Rev. D **29**, 1233 (1984)
[22] P. J. de A. Bicudo and J. E. F. T. Ribeiro, Phys. Rev. D **42**, 1611 (1990).
[23] F. J. Llanes-Estrada and S. R. Cotanch, Phys. Rev. Lett. **84**, 1102 (2000).
[24] F. J. Llanes-Estrada and S. R. Cotanch, Nucl. Phys. A **697**, 303 (2002).
[25] A. V. Nefediev, J. E. F. T. Ribeiro, and A. P. Szczepaniak, JETP Lett. **87**, 271 (2008).
[26] A. P. Szczepaniak and E. S. Swanson, Phys. Rev. D **55**, 1578 (1997).
[27] D. Zwanziger, Phys. Rev. Lett. **90**, 102001 (2003).
[28] J. Greensite and S. Olejnı´k, Phys. Rev. D **67**, 094503 2003!
[29] P. Guo, A. P. Szczepaniak, G. Galat'a, A. Vassallo, and E. Santopinto, Phys. Rev. D **78**, 056003 (2008)
[30] T. Yépez-Martínez, A. P. Szczepaniak and H. Reinhardt, Phys. Rev. D **86**, 076010 (2012).
[31] J. Greensite and A. P. Szczepaniak, Phys. Rev. D 91, (2015) 034503.
[32] J. Greensite and A. P. Szczepaniak, Phys. Rev. D 93, (2016) 074506.
[33] T. Yépez-Martínez, P. O. Hess and O. Civitarese, work in progress to be published.
[34] P.A. Zyla et al. (Particle Data Group), Prog. Theor. Exp. Phys. **2020**, 083C01 (2020).

AUTHOR INDEX